PROTOCOLS, SERVERS AND PROJECTS FOR MULTIMEDIA REALTIME SYSTEMS

Also by Dimitris N. Chorafas

FINANCIAL MODELS AND SIMULATION
HIGH-PERFORMANCE NETWORKS, PERSONAL COMMUNICATIONS AND MOBILE COMPUTING
CHAOS THEORY IN THE FINANCIAL MARKETS
RISK MANAGEMENT WITH OFF-BALANCE SHEET FINANCIAL INSTRUMENTS (*with Heinrich Steinmann*)
INTELLIGENT MULTIMEDIA DATABASES
DATABASE MINING BY FINANCIAL INSTITUTIONS (*with Heinrich Steinmann*)
DERIVATIVE FINANCIAL INSTRUMENTS
MEASURING RETURN ON TECHNOLOGY INVESTMENTS
OBJECT-ORIENTED DATABASES (*with Heinrich Steinmann*)
SOLUTIONS FOR NETWORKED DATABASES: How to Move from Heterogeneous Structures to Federated Concepts (*with Heinrich Steinmann*)
MANUFACTURING DATABASES AND COMPUTER INTEGRATED SYSTEMS
DO IT OR DIE: Database Solutions for Financial Institutions (*with Heinrich Steinmann*)
THE NEW INFORMATION TECHNOLOGIES: A Practitioner's Guide
TREASURY OPERATIONS AND THE FOREIGN EXCHANGE CHALLENGE: A Guide to Risk Management Strategies for World Markets
THE NEW TECHNOLOGY OF FINANCIAL MANAGEMENT
THE GLOBALIZATION OF MONEY AND SECURITIES
SIMULATION, OPTIMIZATION AND EXPERT SYSTEMS: How Technology is Revolutionizing the Way Securities are Analyzed, Underwritten and Traded
EXPERT SYSTEMS IN BANKING (*with Heinrich Steinmann*)
RISK MANAGEMENT IN FINANCIAL INSTITUTIONS
THE HANDBOOK OF MANAGEMENT
SUPERCOMPUTERS (*with Heinrich Steinmann*)
INTELLIGENT NETWORKS (*with Heinrich Steinmann*)
KNOWLEDGE ENGINEERING
NEW TECHNOLOGIES: A Manager's Survival Guide
BANK PROFITABILITY
HANDBOOK OF RELATIONAL DATABASES AND DBMS
IMPLEMENTING NETWORKS IN BANKING AND FINANCIAL SERVICES (*with Heinrich Steinmann*)
MEMBERSHIP OF THE BOARD OF DIRECTORS
ELECTRONIC FUNDS TRANSFER
HIGH TECHNOLOGY AT UBS: For Excellence in Client Service (*with Heinrich Steinmann*)
STRATEGIC PLANNING FOR ELECTRONIC BANKING

Protocols, Servers and Projects for Multimedia Realtime Systems

Dimitris N. Chorafas

Foreword by Eike Jessen

MACMILLAN

First published 1997 by
MACMILLAN PRESS LTD
Houndmills, Basingstoke, Hampshire RG21 6XS
and London
Companies and representatives
throughout the world

ISBN 0–333–66267–9

A catalogue record for this book is available
from the British Library.

This book is printed on paper suitable for recycling and made from fully managed and sustained forest sources.

10 9 8 7 6 5 4 3 2 1
06 05 04 03 02 01 00 99 98 97

Printed in Great Britain by
Antony Rowe Ltd
Chippenham, Wiltshire

Contents

List of Figures

List of Tables

Preface

The subject of new protocols, multimedia servers and leading projects in telecommunications technology is very timely. In February 1995, the *Group of Ten* (G-10) most important industrial countries decided to finance in a big way high-technology in connection to the *information superhighway*, *massively parallel computing* and *advanced multimedia applications* such as virtual reality – all are reflected in this book with examples and first-hand experiences.

- The primary goal of this book is to be inter-disciplinary – hence the range of subjects which it includes.
- Each one of its three parts could stand alone as a separate text. But what is today particularly necessary is to merge the different disciplines into one integrated whole.

The input received in the course of the research meetings which led to this book permits the identification of new domains of applications in communications, an elaboration of the developing implementation perspectives, and a presentation of the breakthroughs which have currently been achieved. It also makes possible the discussion of the most outstanding research projects, and outlines the systems solutions necessary to fruitfully exploit new technology.

Written in a comprehensive and comprehensible manner, this book addresses three populations of readership. The first is senior executives of user organizations and government agencies who are responsible for choosing, authorizing and supervising projects in advanced communications systems – aiming to lead in the marketplace. The text includes case studies on both industrial applications and government-financed projects.

Second, corporate strategists among computers and communications companies – as well as Chief Technology Officers, Vice Presidents of Engineering, Research and Development Directors. Also part of this population are systems architects, network designers and database administrators.

Third, graduate students in colleges and universities. Typically, these will be studying Engineering, Telecommunications, Information

Science, Project Management, Software Development – or work in Media laboratories. All can use this text either for instruction or for reference purposes.

Forty-three years ago I had a professor at the University of California who taught his students that if one truly wants to learn from the experience of others he should not study their successes but their failures. Therefore, I have taken extra care to identify the failures which have taken place on many projects, appreciating that learning from these failures is the best policy for not repeating them in the future.

This book has been designed to lead the reader through the ways and means of communications, computers and software technology, permitting him or her to capitalize from experience so far acquired in complex and expensive network projects, similar to those the Group of Ten is now financing. The material has been divided into three parts:

- *Part One* outlines the technical aspects but also the managerial choices connected to the use of telecommunications networks, and gigastreams of channel capacity.
- The theme of *Part Two* is multimedia, from their business perspectives to the need for specialized servers, and to the projects undertaken in this regard.
- The aim of *Part Three* is experience transfer. It provides the reader with insight and foresight from known projects targetting communications, computers and software which did not go according to plan.

Chapter 1 explains what can be expected from the Information Superhighway after the recent twists in terms of mergers, acquisitions and local agreements, which changed the original projections towards a fairly different orientation. Both the consumer and business viewpoints are explored, and the same is true about the evolving technological landscape because of the effects of deregulation.

The new telecommunications perspectives must be sustained through more efficient protocols and line disciplines than we have used in the past. Chapter 2 explores this subject, particularly from the viewpoint of business opportunities provided by system solutions like the different variants of fast packet-switching methodology.

Chapter 3 focuses on the asynchronous transfer mode (ATM) as a backbone technology but also in terms of integrating local area networks, metropolitan area networks and wide area networks. Chapter 4 compares the synchronous digital hierarchy (SDH) to Sonet and other implementation protocols for telecommunications. This completes Part One, which is the section of the book addressing the new telecommunications technology.

The business perspective of interactive multimedia as well as the race for multimedia servers and their implementation is described in Part Two. Chapter 5 covers the technical and technico-economic aspects of multimedia. It also includes case studies from the business environment, where lies the greater market potential.

The different projects currently focusing on multimedia servers are described in Chapter 6, which also includes issues such as image modelling, digital preassembly, as well as the synergy between multimedia servers and the asynchronous transfer mode. This leads nicely to the practical applications of multimedia and virtual reality, which constitutes the theme of Chapter 7.

Chapter 8 brings to the reader's attention broader projects such as GreenSpace. Good lessons can be learned from the fact that the five phases of Project GreenSpace underline issues of network infrastructure, the new terminals to handle wideband input/output, and other challenges like intentional and extensional database solutions.

Part Three looks at the way pioneering projects are being managed. The focal point of Chapter 9 is the American High-Performance Communications and Computers (HPCC) program, its achievements and its shortcomings – as well as results obtained from other less lavishly financed projects, which can help as examples on strengths and weaknesses.

Chapter 10 explains why research and implementation projects in telecommunications need parallel computer programming. It takes as an example the Japanese Advanced Telecommunications Research (ATR) laboratory and marries some of its requirements with a parallel programming project at the University of Virginia.

To complete the subject of what the taxpayer gets back from government-financed projects in communications, computers and software, Chapter 11 looks at return on investment in connection to HPCC. It also examines the (meagre) outcome of ICOT, a Japanese government-financed project on parallel computers – as well as taking a look at the European Union's High-Performance Computers and Networks (HPCN) programme. All these projects have one thing in common: it is easier to develop new technology than to develop new attitudes – yet the latter are a prerequisite to the successful application of the former.

* * *

I wish to express my appreciation for the collaboration received from some 300 senior executives, systems designers, computers and communications specialists and other experts in America, England, Germany, Japan and many other countries.

Ninety-four computer manufacturers, communications companies, financial institutions, service companies and university laboratories participated in this effort. A personalized list can be found in the Acknowledgements. I would like to particularly underline my appreciation to Dr. Heinrich Steinmann, Dr. Alan W. Rudge and Professor Eike Jessen.

It is also appropriate to express my thanks to everybody who contributed to this book: to my colleagues for their insight, to the company executives and technologists as well as the university facilities for their collaboration, to Keith Povey for copy-editing and to Eva-Maria Binder for the drawings, typing and index.

Valmer and Vitznau DIMITRIS N. CHORAFAS

Foreword

The integration of our computing and data resources by high-performance networks is an evolution able to change our forms of life, society and economy to as large an extent as it has been with the evolution of the transport of persons and goods in preceding centuries. Our own individual existence – as information and society-dependent beings – is as much affected as that of the organizations we have created for our purposes: states, enterprises, even our knowledge structures and processes.

For our current orientation and midterm strategy, the boundary conditions imposed today by technology, procedures, and organizational structures are stringent. A hind-sight study is often more enlightening for the future than prophecies. In a rapidly accelerating process as that of the integration of computing and communication, many enterprises find themselves in a decision situation the outcome of which can be very critical for the existence of the enterprise.

- In this situation, the potential of technology must be conceived by top management.
- Vice versa, its technologists must learn and respect the needs of organizations and enterprises, or both will be the losers in the end.

Technology is not only hardware and basic software implementing functions on networks, user access systems and server computers. It is a world of policies and procedures and (partially standardized) services that will provide an integrated but discrete, an open but safe, a distributed but consistent information world. Most of the open challenges, in particular those to be met by a joint effort of management, end users, and technologists, are here.

Only in low level issues in the computers and communications world, do there exist standards. Ethernet, X.25/IP networking, maybe ISDN, are examples. The victory of open systems over the closed proprietary systems (though functionally often superior) is a lasting event. Internet as an open network, though initially inacceptable in principle, proved to be incredibly effective. But for most of the higher levels there are still wide uncertainties.

The careful consideration of advanced pilot projects and of research projects is important for the determination of our strategy. Here, the impact of advanced multimedia user interfaces on data structures, algorithms and communication demands has a striking effect on the usage of distributed systems. World Wide Web is just another example of a research spin-off with an amazing success.

The new information processes and systems will widely change the tools and the environment in which managers and professionals have to prove their ability. Banking, insurance, transport enterprise and educational institutes are most sensibly dependent on the usage of the new systems.

It is the unique merit of this book to put its reader in the focus of the technical prospects and the organizational needs. It has been written out of thorough experience with both aspects of the problems which lie ahead. Therefore it addresses technologists as well as management and gives them that information and insight which permit them to get through the very critical learning and decision process on which the future of our economy will depend.

Professor Dr EIKE JESSEN
Vice-Chairman, Technical
University of Munich

Part One

How to Use Telecommunications Protocols and Gigastreams

1 What's the Sense of the Information Superhighway?

INTRODUCTION

The information revolution of the 1990s began in the United States, and while it is gradually spreading to other countries this diffusion is uneven. Many criteria make this statement valid: from cabled cities and towns to the national initiative taken with the goal of bringing the country into the 21st century.

In America and Canada, coaxial cable passes by more than 80 per cent of homes – up to 95 per cent in parts of Canada. Switzerland also has a dense cable TV (CATV) network and now lays an optical fibre net – as a result of competition in cable television, but not yet in phone services.

In England, a 10-year-old deregulation sees to it that competition in telephony has picked up. But in the rest of Europe a slow deregulation delays the introduction of new competitors into telecommunications and cable TV.

Another criterion on how well are laid the tracks of the *Information Superhighway* is home computers. In Germany, more than one-fourth of all homes are PC-equipped – and the same is true in the United Kingdom.

In the Pacific Rim, Australia has the largest per capita installed base of PCs: one for every 5 people, compared with one for every 20 in Japan. But cable TV and multimedia are more advanced in Japan than in Australia.

In South Asia, India's industrial base and large middle class make it a natural candidate for technological advances. But the country has been very slow in moving in the computers and communications (C&C) direction. The impedance is the government, which for several decades was feet-dragging in computers and it has only recently and timidly begun to deregulate telecommunications.

In Latin America, Argentina, Brazil and Mexico have fair but not advanced telecom infrastructures. Some countries, like Brazil, overprotected their local computer industry with the result of falling way behind in technology. Latin America lags quite significantly in C&C.

- Mexico has one PC for every 50 people, and
- Only half of all Brazilian homes have telephones.

As for the black continent, only South Africa has the capacity to build high-speed telecommunications networks within a reasonable time. Similarly in the Middle East, Israeli technology is way ahead that of its neighbours – and by every prognosis it will continue being that way well after the year 2000.

Therefore when we talk of the landscape in which information superhighways can be built with good chances of success in terms of usage and return on investment, the references made mainly concern North America, the European Union, Japan, a few other countries in the Pacific Rim, as well as Israel and maybe South Africa.

THE SEARCH FOR LEADERSHIP IN INFORMATION SUPERHIGHWAYS VENTURES

In the fifty years which have followed World War II, our society changed from manufacturing-intense to service-oriented, just as in the interwar years the change was from agricultural to manufacturing. But there is also a cultural revolution which particularly characterizes the last 20 years: we have become a *multimedia* society. This has three aftermaths:

1. There is a realtime diffusion of news around the globe, with the predominant effect being visual.
2. Therefore many, too many, people tend to think with their eyes rather than their brains, and
3. News carried by media is often taken as representing the real thing – though there may be no completed deals to substantiate such statements.

The net result of this happening is that big plans may make news but the deals that are supposed to take place are not always completed. While different reasons may be in the background of engagements which are broken up, when break-ups become a trend it is necessary to revise predictions.

This statement is just as valid for the projected information superhighway which has had its share of plan reversals. While the infrastructure of a knowledge society will eventually become a reality, *when* exactly this will happen and *which* precise form it will take, are not yet 'sure things'. Among the better-known deals-to-be which are now dead can be found:

1. *Microsoft/Time Warner/Tele-Communications (TCI).* Talks on a joint venture in technology started in June 1993 – but led nowhere.

Multiparty, big technology deals are always hard to pull off and it seems that there have also been cultural conflicts. In the background of this effort was the drive to set standards for the transmission of interactive programmes. The deal-to-be fell apart after a month of negotiations.

2. *Bell Atlantic/Tele-Communications*, a $21.4 billion projected merger, which came online in October 1993 – then faded.

The two parties blamed this deal's demise on several uncertainties raised by new cable regulations that could hurt profits and stunt growth. But industry analysts debited the failure to the clash of TCI's rough-and-tumble ways with the staid Bell Atlantic culture.

3. *Southwestern Bell/Cox Enterprises.* Negotiations on this $4.9 billion joint venture begun in December 1993, then died out.

As in the case of Bell Atlantic/TCI, it was said that new cable regulations caused Southwestern Bell to drop the initiative. Wall Street analysts said that, rather, the flaw was in the idea of a marriage between a Baby Bell and a CATV operator – which may bode ill for other deals of that kind.

4. *Electronic Data Systems (EDS)/Sprint.* This has been a projected merger creating a $20 billion company, which in the end did not flourish.

Discussions between Sprint and EDS started in May 1994, but talks were suspended a month down the line. This happened after a dispute over how to value respective operations. By all likelihood, this will be a recurrent theme in Information Superhighway-type mergers.

The irony of the breakdown in the EDS/Sprint project is that bringing together the two companies could have made sense, because it offered Sprint software expertise while EDS could benefit from Sprint's experience in running big telephone networks. At least by one estimate, more than a third of the fees paid by many EDS's customers for the management of their computer networks goes for telecommunications.

- At present, EDS subcontracts such telecoms work to a variety of telephone companies.
- A merger with Sprint could have brought both skills and revenues in-house.

EDS was also interested in gaining access to Sprint's 8 million long-distance customers and to the fully digital nationwide fibre-optic network Sprint has built. This was also an important asset for General Motors, the parent company of EDS.

But not everything is negative in terms of deals which can make the information superhighway a reality. Other deals were completed such as Time Warner/Toshiba/Itochu – a $500 million investment. In October 1991 the two Japanese companies together acquired 12.5 per cent of Time Warner Entertainment which includes Time Warner's film studio, cable systems and Home Box Office.

Nearly a couple of years after that agreement, in May 1993, came the Time Warner/US West deal, which represented a $2.5 billion investment. US West got 25 per cent of Time Warner Entertainment. Then, October 1993 saw the Nynex/Viacom deal, amounting to a $1.2 billion investment. Nynex bought preferred stock in Viacom to help in its battle for Paramount.

The biggest deal in the 1993–94 timeframe, which started August 1993 and was completed about a year later, has been AT&T/McCaw Cellular Communications; a $12.6 billion acquisition. In 1994, another information superhighway-type deal has been that of Bell Canada Enterprises/Jones Intercable; a $400 million investment. Then, in mid-1995 Disney merged with Capital Cities/ABC. While the resulting company's activities span from programming to distribution, management seems to appreciate that to handle such a complex project successfully it has to dot the i's and cross the t's of the new venture. This job still has to be done.

There have also been other deals of which we will talk to a greater extent in Part Three. For instance, the joint venture of MCI and Murdoch's News Corporation; as well as the Time Warner–Turner merger which made, for the time being, the largest media corporation in the world.

If we compare the deals which have been done to those which failed, we will see that the former are more homogeneous than the latter – which may be a reason why they were successful. The query, however, remains of whether size really provides economies of scale or, to the contrary, swamps productivity. If the merger of Chase Manhattan and Chemical Banking is taken as an example, this rush to the altar may be an effort to hide weaknesses rather than to amplify strengths.

Finally, if we focus on telecommunications firms rather than media companies, business-type considerations are very important because the megastream and gigastream networks (of which we will talk in Chapter 2), and the new, efficient protocols (covered in Chapters 3 and 4) must demonstrate that they have:

- A reason for existence, and
- A steady income basis.

Down to basics, the reason for existence is the *virtual corporation* and the *virtual marketplace* whose nature and projected activities are discussed in following sections. In terms of products, a great deal of the income stream will probably come from *multimedia applications*. What this means, as well as the requirements which it imposes, will be examined in detail in Part Two.

With the exception of AT&T/McCaw, none of the investments or acquisitions which have been completed so far approach the scope and magnitude of the four which died. In terms of mergers or megadeals, the information superhighway has started on a bumpy road. However, nothing is steady in business and things may change in the years ahead.

WHAT CAN BE LEARNED FROM COMPLETED AGREEMENTS AND BROKEN DEALS?

A couple of busted cable-TV/phone company deals-to-be will not kill off the information superhighway. It is going to happen but it will take place slower than originally thought, as some companies start to realize that they probably do not have to spend so much money to compete in the new telecommunications environment.

We also said in the previous section that not all cable–phone deals went bust. AT&T remains confident about the future of its $12.6 billion acquisition of McCaw Cellular Communications. In fact, after Federal District Judge Harold H. Greene stopped short of approving it in April 1994, both companies seemed willing to tinker with the deal to win over approval.

By contrast, on the broken deals side Bell Atlantic and Southwestern Bell opted against restructuring their original plans to account for reduced cash flow by TCI and Cox. Wall Street analysts believe that the Baby Bells were less willing to compromise because they saw that cable companies were not necessarily essential in offering broadband video services.

Pacific Telesis, for example, decided to go it alone by laying optical fibre in California. And Bell Atlantic entered into a deal with Oracle and nCube to build multimedia servers to its specifications – a strategy also followed by British Telecom, which has a similar project with nCube and Oracle.

Not everything is, however, in cable and fibres. Programming is a skill that cable operators are supposed to provide to a cable–phone partnership. As we have seen, in 1993 US West payed $2.5 billion for 25.5 per cent of Time Warner Entertainment. Included in the assets are Time Warner's cable channels and movie studios.

On the marketing side, an obvious target for US West is New York City, which is part of Nynex's territory. In Manhattan, Time Warner has been operating for about 20 years a cable system serving a million subscribers. Correctly, the strategy of US West is to go after a solid customer base:

- Adding to fibre-optic cable that Time Warner has already installed, the telco will piece together a ring able to serve major businesses in Manhattan.
- Each customer will have a line from his office building to the ring, able to take calls to and from long-distance companies.

Eventually, US West would help its cable partner extend its fibre lines into the neighbourhoods, where they carry both video and phone traffic to all of Time Warner's subscribers. In fact, for any practical purpose this is the strategy underpinning all cable–phone agreements.

As this and other references help to document, it is not difficult to state the hypothesis that rather than being set back the Information Superhighway is really taking shape. The lanes that will help companies conduct their business electronically are being laid, and the management of tier-1 firms recognizes that:

- A high-speed information infrastructure can play a major role in speeding the pace of operations and cutting costs.
- Intelligent broadband networks have the potential to boost responsiveness and save money every time a transaction can be changed from paper to online communications.

Business worldwide is also under constant pressure to dismantle outdated vertical organization structures and to intensify partnerships with other companies. The drive is towards *virtual companies* which use *virtual offices* and employ virtual reality for teleconferencing (see also Chorafas and Steinmann, 1995).

Today, whether in finance, manufacturing or merchandising, the most effective way to manage corporate resources is through *federated* independent business units. What is meant by federalism is not just another type of decentralization. The conceptual differences are important.

- *Decentralization* implies that the centre *delegates* certain tasks or duties to the outlying units, while the centre itself remains in overall control.

With this model, the centre does the delegating, the initiating and the directing. The business organization is still hierarchical, though not everything is decided at the centre.

- *Federalism* is different. The centre's powers are given to it by the outlying groups, in a sort of *reverse delegation*.

The centre, therefore, coordinates, advises, influences and suggests – but it does not order. Multimedia and the broadband communications capacity of the information superhighway are in particular

Figure 1.1 Seventy years of evolution in the dimensions of industrial organization and structure

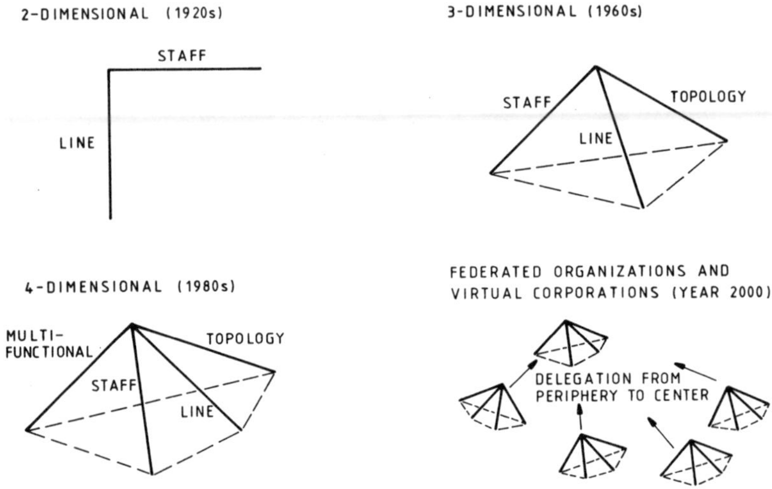

demand with this organizational model – but let us recall that neither in a structural sense nor in technology has the federalist solution developed overnight. Instead, the evolution has been gradual.

Figure 1.1 presents four facets of the development which took place in industrial organization over 70 years. The milestones are the hierarchical staff and line; the equally hierarchical 3-dimensional aspect of staff, line and topology; the evolution of this model towards the decentralized organization with a 4-dimensional chart; and the n-dimensional federated business organization which will characterize virtual corporations.

PAYING ATTENTION TO THE NEEDS OF THE VIRTUAL CORPORATION

The gradual evolution towards a new type of business organization has had a multiple background. The 2-dimensional structure of the organization chart – which long ago became a classic – has been the model Alfred Sloan developed in the 1920s at General Motors.

But the international business expansion twisted this model.

After World War II, transnational operations brought a third dimension to the chart, which could not be effectively represented on a piece of paper. Neither could plain old telephone lines serve business management needs anymore. Conglomerates saw to it that the discrepency between what was needed and what was done increased:

- From product planning to the analysis of profit and loss, a fourth dimension has been introduced.
- Sloan's staff-and-line model could no longer effectively serve; a more polyvalent approach became necessary.

As a result, the first concept to be rejected has been the highly centralized, hierarchical organization. It has been put to rest because it proved to be ineffectual, and it has been replaced by a decentralized or divisionalized approach.

Decentralization distributed not only product and market responsibility but also the information technology supports to the divisions. But it did not change the tools, which was a mistake. Up to a point, this remained true with the federated organization in the sense of independent business units (IBU) which, in an organizational sense is 4-dimensional.

Federalism is how the Board of Management works in central European companies. The members of the Board are thought to be equal or semi-equal. The chairman practically acts as the speaker of the Board, reigning by consensus. There is no chief executive officer with 100-per cent control.

In the transnational corporation, an IBU is the smallest identifiable hole within the larger aggregate. It has research, development, manufacturing and marketing under its jurisdiction but auditing and control is exercised by the centre preferably online in realtime – to assure that:

- Exposure is kept under lock and key,
- Concurrent business activities are effectively carried out, and
- No centrifugal forces work to break up the corporation.

Independent business units are responsible to the larger organization in terms of profit and loss (P&L). At the same time, they have their own management and act autonomously in terms of business transactions which interest them locally, yet they are always ready to collaborate in global events. Virtual corporations are an extension of this notion:

- The concept of 'less than 100 per cent control' is bread and butter among the foremost companies.
- It is a powerful approach which merits a great deal of attention to be paid to the essentials, rather than to micromanagement.

Fast software development, agents (proactive knowledge artefacts), desk area networks and broadband networks are the enabling technology for virtual corporations. Therefore, the concept of an information superhighway has an ample applications domain in industry. Eventually, the same will be said about households.

Intelligent multimedia networks play a significant role in connection to a federated business and industrial organization's ability to survive. A modern company's business success is directly linked to its ability to bring high-quality products and services to market faster and more cost-effectively than its competitors. This requires a whole new approach to:

- Working concurrently on design and engineering projects
- Developing and maintaining an interdisciplinary initiative by utilizing the most advanced technology, and
- Addressing client demands for customization in geographically distributed markets

Satisfying the needs and expectations of the customer, both those expressed and those implied, provides significant competitive leverage in the current global, dynamic but also turbulent business environment.

Global competition, increasingly fragmented markets, more knowledgeable and more demanding end-user companies or individuals, and shrinking product life cycles tend to escalate the costs of development. Short time-to-market and affordable costs require interactive design technologies, intelligent computer networks, desktop video-conferencing, and rapid prototyping tools.

NETWORKS TO SUPPORT THE VIRTUAL MARKET-PLACE

Part of the culture of the virtual corporation is the virtual market-place. One of the first examples is provided by a group of 20 big Silicon Valley companies, including Intel, Sun Microsystems, Hewlett-Packard, Pacific Bell and Apple Computer which, in April 1994, jump-started the most ambitious electronic market-place so far. This initiative is backed by $12 million in federal, state and private funds for developing the needed software.

- *CommerceNet* is designed for more than getting purchase orders, bills and other similar documents moving online.
- The main goal is to serve as a technological platform in which to test all sorts of ideas for conducting commerce over the Internet and other data networks.

It is expected that CommerceNet will inspire other industries to build similar electronic market-places, as well as help in shaping America's electronic infrastructure. This is Information Superhighway stuff which goes well beyond what some other initiatives undertake, such as digital access to libraries.

While it has its technical merits, digital access to existing libraries is not an exciting subject; some experts even considered it to be a rear-view-mirror concept. We should think ahead about more imaginative uses of new resources, which could radically improve the way we live and work – for instance, education.

One way of looking at the information superhighway is as a catalyst. New educational theories may help to guide teachers and students into an interactive educational technology. Similarly, CommerceNet might help to produce an interactive electronic market-place:

- A starting point may be task analysis able to identify primary and secondary services for the customization of a mass market.
- Based on personalization requirement of the client base systems designers can prepare prototypes to validate concepts.

For their part marketing experts can elaborate a methodology to assure consistency in terms of sales efforts and transactions. The electronic market-place may help to significantly shorten the time to market. Toyota projects that it will be able to deliver a custom-made car in 3½ days, by the end of this century.

If CommerceNet works as planned, its collection of independent but closely interconnected computers will evolve to support a variety of electronic links between businesses. The goal is to help different companies work together as a virtual corporation.

- Intelligent agent programs have been planned for delivery, to assist in this initiative.
- The first lot of these would ferret out the best deals in microchips and other parts in the electronic market.

One of the big payoffs is expected to be in more efficient methods for companies to distribute product information. Hewlett-Packard, for instance, reckons that it currently keeps from 30 000 to 40 000 printed items on hand to describe its 20 000 different offerings.

The change from paper to information held and handled by distributed communicating databases will be significant. Since these databases are accessed online, buyers will be able to interactively pore over descriptive references, tolerances, costs and other details electronically, browsing as they search for target pages and viewing volumes of multimedia catalogues that may feature:

- Voice animation, and
- A growing array of video clips

As this and similar efforts help to demonstrate, business and industry currently face the prospect of a profound change in information handling. Instead of data on microcomputers, which characterized technology in the 1980s, a new era of multimedia telecommunications opens up – based on the convergence of micros with communications networks.

This landscape is scattered with all kinds of opportunities to the leaders, as well as risks to the laggards. Therefore, as we saw in the preceding sections, telephone companies are buying into programming outfits and computer companies are joining with telephone and cable companies in order to develop better rounded, appealing products.

- Everybody's goal is to carve out a place in an *all-digital future.*

- This is where the markets of the late 1990s and well into the 21st century are expected to be.

Clearly, 1995 marks the beginning of a new era in communications and computing, one in which increasingly sophisticated software plays a major role. Businesses compete for world domination of various communications, computing and software technologies. New products will emerge in the coming years and will flourish, if deregulation and legislative action fertilize the field in which they are planted.

EFFECTS OF DEREGULATION AND MARKET ACTION

Information superhighway services can be both used and misused. New product lines sometimes spring out of the blue, propelled by fashion. An estimated 1.5 per cent of the world's international telephone calls are now placed to offshore phone-sex services. Americans are dialing up 15 million minutes a month for dirty talk from overseas, according to a study by Tele-Geography (1994).

- That makes the US the Number 2 foreign phone-sex user, behind Germany.
- But also in France the most popular product on Minitel, the national videotex, is said to be sex-related services.

There have been several discussions on whether the use of the new communications facilities should be regulated. They ended in no consensus, but by far the prevailing opinion is that some misuse is inevitable. Regulators and legislators should not repeat mistakes of the past.

In a climate of increasing international political and social change, any economic activity, such as a new plant for telecommunications services, has an impact on existing power balances. In fact, decisions related to communications, computers and software take on cultural and political dimensions to which regulators and legislators must be very sensitive.

Intelligent Networks, a book I co-authored in 1990 with Dr Heinrich Steinmann, made the forecast that the barriers dividing the three areas of authority of the Federal Communications Commission will

Figure 1.2 The integration of three areas of activity supervised by the Federal Communications Commission

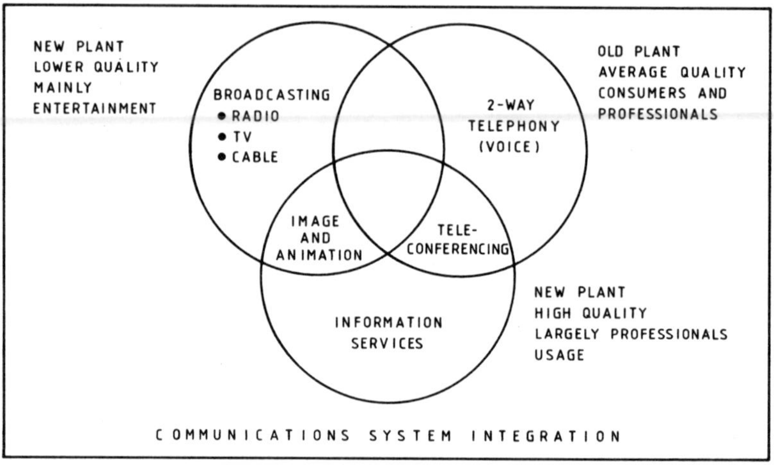

most likely fall over a five-year period. This took place, and Figure 1.2 identifies the integrated domain which has since taken shape – with a huge impact on economic activity. (Another prediction made in 1990 was that this integration would wash away ISDN – which has also been the case.)

Projections made today about telecommunications in the year 2000 and beyond usually assume a continuity in government policies and business strategies. One has to be careful about such hypotheses because time and again they have proved to be wrong. Taking the United States as an example, during the 20th century there have existed four periods of discontinuity in the telephone infrastructure:

1. Just before World War I, when the US government nationalized Ma Bell.
2. Fifteen years later, in the 1920s, when AT&T was reprivatized.
3. In the early 1930s, when the US government decreed that competition within the telephone industry was detrimental to the national interests, and
4. In the mid-1980s, when the US government decreed that lack of

competition within the telephone industry was detrimental to national interests.

Decision No. 1 gave AT&T lots of perks and privileges which it largely retained after decision No. 2. Decision No. 3 promoted a monopoly and, with it, a slow-moving telephone bureaucracy like the European PTTs. Decision No. 4 has been the decree which led to deregulation.

When, in 1984, America deregulated its telephone industry, most economists predicted a market boom, caused by competition, innovation and technology. This happened. What the economists failed to predict, however, is that a regulated monopoly would turn into a deregulated oligopoly.

One of the items now under discussion is that the liberalization of telephony was imperfect because deregulators did not understand the forces at work during regulation. In the background of this assertion lies the fact that:

- Both a market and the systems infrastructure serving it consist of a family of products and processes.
- Each is composed of dynamic parts which interact with and often displace each other.

If we look back some 130 years at the origin of the telephone service, we see that the telephone displaced the telegraph as the communications medium. Earlier on, the telegraph had taken over from messages carried by pigeons and by horse and carriage.

Part of the success of telephony is due to its capillarity. Another part is credited to the interactive, 2-way message exchange. But more recently, some of the 2-way voice communications have been replaced by electronic mail and voice mail.

In the 19th century, the postal service replaced the mail carried by horse and carriage, but the real boom in the mail business came after the standard stamp (as we know it) was introduced in England towards the end of the last century. More recently, as the postal system became clogged with garbage mail, courier service took off – while facsimile, long dormant but revived by newer technology, took away another big chunk of postal business.

Market action often precedes both regulation and its antithesis: deregulation. Quite often, legislators act only when there is no other alternative because the wave of change has greatly altered the landscape which was familiar to them. This does not happen only with telecommunications, but also with many other products – derivative financial instruments being an example (see for example Chorafas, 1995, and Chorafas and Steinmann, 1994).

VALUE DIFFERENTIATION THROUGH INTELLIGENT NETWORKS

The implementation and operation of a new system, like the intelligent, multimedia broadband networks we call information superhighways, can be successful only when they gain market support. Experience shows that this is done by way of new products.

New products, such as live videoconferencing and virtual reality (discussed in Chapter 7) are now appearing and heading towards market acceptance. By the year 2000, a large number of telecommunications-related technologies will require broadband services to survive and grow. This will not happen in a linear way, but it will be characterized by market advances and a number of retreats, like the broken deals we saw earlier.

Looking at the information superhighway from an infrastructural viewpoint, the whole concept rests on four pillars which are identified in Figure 1.3. Let us notice, however, that these are essentially the pillars supporting all communications having to do with information:

- Generation,
- Transfer,
- Storage, and
- Usage.

The way to bet is that every one of the developments connected to this frame of reference will, in all likelihood, be nonlinear. If business moved in a predictable and linear way without discontinuities, we could easily foretell what would happen next – which is not the case.

But some issues are indeed predictable. For instance, it is fairly sure that in the coming years companies will be increasingly using their network as an internal tool to boost the efficiency of their

Figure 1.3 The four pillars on which modern communications rest

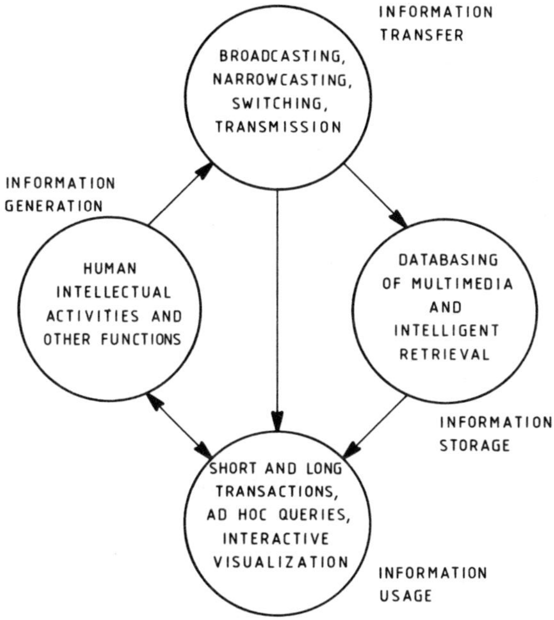

operations: from design engineering, to manufacturing plants and the sales force.

- Although averaging over 50 hours per week on the job, each salesperson today invests only 15 to 20 hours with potential buyers.
- The rest of the time is usually spent preparing bids, travelling to the prospect's premises, getting training, being updated on new products as well as the competition, and so on.

The evident goal is to use telecommunications to significantly raise the time spent with customers. The virtual office has this goal, and if this effort is successful, as looks likely, it will present some interesting discontinuities in current projections – by creating new problems and new opportunities of which we are not currently aware.

The message to be retained from this reference is that the world economy is changing fast, and such change shows signs of becoming even faster – as well as global. In the Pacific basin, for example, a new focus of economic activities is emerging; the US is rapidly restructuring itself into a post-industrial economy; and in Europe the liberalization in Eastern Europe has altered the market dynamics.

Demographic changes create new challenges whose aftermaths are still unclear. What is sure is that they are not linear. At the same time, the new international division of labour has a great deal to do with fast-developing technologies – the two together setting the scene in a discontinuous way.

One thing we could say with a reasonable level of assurance is that in the years ahead information will be one of the most important basic resources – together with raw materials, environmental concerns, skilled labour and energy supply. The capture, transmission, storage and processing of information will:

- Determine and control future industrial and economic development, and
- Have a major impact on everyday life because it influences so many consumers.

But are we ready to manage the new information-intense environment? Are we able to look beyond our enthrallment with technology and alluring commercial promise of 'this' or 'that' service or device? Are we able to explain to likely users exactly what the information superhighway is, how it will work and what its probable effects will be?

As an intelligent network, the information superhighway will be an integrated multimedia system supported by knowledge artefacts, using terrestrial (fibre and coaxial cable) channels as well as radio links and satellites.

- It will be composed of many heterogeneous networks.
- Users can interconnect through these with multipurpose tools.

If there is no monopoly or oligopoly connected to the information superhighway, which is not yet a sure thing, then the new structures will develop into very intelligent carriers. User organizations and private individuals will be able to attach knowledge-enriched devices:

- Some polyvalent, which can do many things their masters want them to,
- Other specialized, designed for a single purpose.

Which way the information superhighway will be structured, operated and marketed is still an open question. The same is true of its reliability. In the past, in connection to integrated networks, early users reported technical problems. These included complete network failure, low network security (a good example is being provided these days with the Internet) and so on.

Security problems repeat themselves time and again, and the same is true of underutilized resources because of lack of skill and too much traditional thinking. Unless these problems are solved there will be no major investments in information superhighway services – and the project will simply falter.

WINNERS AND LOSERS ARE DIFFERENTIATED BY THE COMPUTER CONNECTION

The capacity available in an information superhighway will be used both by companies and by consumers. Financial institutions, manufacturing concerns and merchandising firms are already starting to require gigastreams in channel capacity. Because of the benefits to be obtained, business and industry will be more prone to pay the costs than the larger but economically weaker consumer population.

We have spoken of virtual reality and live teleconferencing, as contrasted to frame freeze, being the new market for gigastream capacity. But there will also exist a number of revamped old markets, an example being database-to-database, computer-to-computer connectivity using the services provided by the information superhighway.

This area of applications is haunted by problems because of mistakes of the past. In corporate computing, the mainframe is a symbol of a bygone era. Cobol programs and mainframes are a legacy of the days when data processing strictly controlled access to computers, the latter being housed in big hospital-looking glass houses.

- Networks of low-cost networked personal computers brought renewal to information processing in the late 1980s, putting computers at the command of individuals, and

- As many companies are finding out, this effect has been amplified with client–server solutions which present significant communications requirements.

Open systems and client–server computing have achieved much lower costs and greater efficiencies than centralized systems. User organizations which stayed with mainframes face continuing problems connected to incompatibility between computers, difficulties in keeping track of corporate data flows, and rising costs for systems management.

In my judgment, the companies able to make the most out of gigastream channel capacity and intelligent networking facilities are those which start from client–server solutions, not the mainframers. Workstations and server vendors, basic software outfits and user organizations are preparing to compete with each other in the markets opened by the information superhighway.

- The main thrust among the foremost companies is agile and polyvalent business-to-business links.
- The goal is to help independent business units connect among themselves, their customers and suppliers – in short, their business partners.

By extending internal networks over high-speed phone lines, companies are aiming to widen the enterprise frontiers. This permits them to communicate and transact business information more effectively. But there are also laggards, particularly the mainframers who have not yet got their signals straight and are therefore missing the advantages provided by the new environment.

IBM provides a good example of what is meant by this failure. Nothing in Big Blue's portfolio of businesses and products can quickly pull the company out of the morass into which its huge computer hardware business has sunk. In the opinion of Wall Street analysts, the problem looks fairly simple.

- Despite efforts to diversify, 49 per cent of IBM's revenue still comes from computer hardware, and
- Virtually every IBM hardware market is being eroded in terms of prices and profits by fast-moving competitors.

As far as Big Blue is concerned, between 1990 and 1994, a total of $19 billion in hardware profits simply evaporated. And there are no signs of a profit rebound. Rather, there is an alarming slide underway in the core mainframe business – the products that used to deliver the highest profits and secured the computer manufacturer's hold on corporate customers.

Worst of all, the company's culture remains for over 30 years pinned to the mainframe size and immobility. Not only was this its core business since it was able to liberate itself from electronic accounting machines (EAM) in the early 1960s, but also mainframes generated revenues from software, leasing, maintenance and other services. The surprise is that immobility still characterizes the once mighty dynamic company.

The reason why I insist on these points is that they describe the *counterculture* to the information superhighway and what it stands for. Even minicomputers, which have been delivering the biggest chunk of hardware profits in the early 1990s became a trouble spot, further hitting the bottomline.

Revenues from AS/400 minicomputers continue to drop significantly, and it changes nothing that IBM blames such a dip on the slow European economy of 1993–94 and its aftermath. The real reason is that American business has moved away from the AS/400 since 1991–92, and European business follows on the same course. Financial analysts believe that minicomputers are just going off a cliff the way mainframe sales have met their doom.

What may be the connection of all these references to the future information superhighway? In 1953, at the University of California, I had a professor of Business Administration who taught his students that one must always bet on the fast-moving companies for:

- Investing,
- Promoting, and
- Utilizing

new products. In a similar manner, Dr Harold Koontz said, if one had to do a tough, time-consuming job and required partners, he should choose people who are very busy rather than those who have little to do. A busy person will always find the time if he really wants to accomplish something. An inactive person will always be short of time and of will.

A company which is moving ahead fast, exploiting high technology and carefully watching its competitive position, is the best partner for the information superhighway. Companies which have fallen way behind have nothing exciting to offer and nothing really to gain from the new high-bandwidth intelligent communications technology.

There is also the problem of government-sponsored research, and along with this the answer to the query whether taxpayer's money is manna from heaven or a hindrance. This subject is vast and controversial, but it is also very important. For this reason three chapters are dedicated to it – which the reader will find in Part Three.

2 Telecommunications Protocols to Fulfill Business Opportunities

INTRODUCTION

Because, as we saw in Chapter 1, from the virtual corporation to the virtual market-place the business opportunities of telecommunications networks are expanding, designing reliable communication protocols is becoming an increasingly complex and challenging task. They need to be not only reliable but also efficient – hence, instrumental in servicing a growing number of distributed, realtime mission-critical applications.

A communications protocol is said to be stabilizing if, starting from or being at any destabilized (illegal) state, the protocol will eventually reach a consistent global state, and resume its normal execution. To achieve stabilization the protocol must be dependable, and able to detect an error when it occurs. Then, it must recover from that error and revert to a legal (stabilized) state.

- These preoccupations were practically non-existent with older communications protocols, including X.25, binary synchronous communications (BSC) and start/stop.
- But the range of diverse and demanding implementations we saw in Chapter 1, from the information superhighway to multimedia applications, has increased the requirements a protocol needs to fulfill.

To appreciate the impact of this comment it pays to look briefly back for three decades, when data transmission came into perspective. During the 1960s and over the subsequent years, forms, formats and protocols developed like wild cacti.

- Not only was normalization not the rule,
- But also each one of the dominant vendors used its own protocols to lock-in the client base.

It is therefore not at all surprising that every major user of computers and communications is today faced with a variety of format and protocol versions, which results both in confusion and in considerable expense since it is necessary to make such heterogeneity transparent to the user.

This chapter has two goals: the most important is to introduce the reader to *fast packet switching* which by all evidence will characterize the information superhighway. The other goal is to do so by bringing to the reader's attention successes, failures and puzzles of the past which by all evidence will present themselves under different forms with the new solutions.

OPEN, EFFICIENT AND SECURE TELECOMMUNICATIONS SOLUTIONS

Which are our alternatives in formatting messages? To be able to answer this query in a valid manner, we must first consider a slightly different question: 'Can we separate *format* from *application*?'

Ideally the answer should have been 'Yes' – and technologically it might be so. But in the real world many facts are forcing the network designer's hand. The query then becomes: 'Can we support with one and only one format database-to-database communications with our business partners – suppliers and customers – now and in the future?' To this, the answer has to be: 'No!'

What this 'No!' answer implies is that in a practical sense the response to the query on whether we can effectively separate *Format* from *Application* also has to be negative because of two reasons: the one is mainly *technical*; the other has to do with *business* perspectives. In the background of the technically-oriented answer is format mapping in an environment where the supports already are:

- Physically different, and
- Logically dissimilar.

Down to its fundamentals, a protocol is a *formalism* which in a modern implementation environment has to address different but complementary functions as shown in Figure 2.1. The one most generally recognized is that of a line discipline, but other functions too are important. For instance, database access mechanism(s) and coding characteristics.

Figure 2.1 In a modern implementation environment protocols have to address different complementary functions

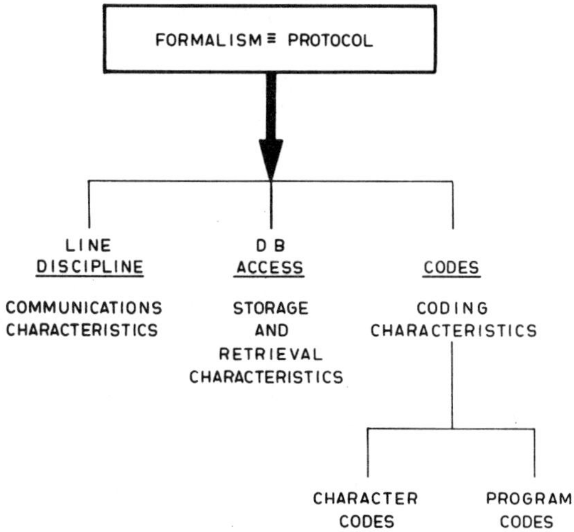

These coding characteristics can be interpreted as being part of the interface between the physical layer, which in the ISO/OSI model is the lower level of reference, and the logical layers leading to the human component (see also references to GUI later in this chapter). A great deal of the challenge is that:

- Investments made over the years have not paid attention to homogeneity,
- Therefore, different heterogeneous solutions exist layer by layer.

No major company today starts its telecommunications solutions from scratch, and this is just as true of most of the minors. Practically all organizations are burdened with systems which are a mosaic of formats and protocols – all the way from the line discipline to database interactivity.

At the same time, in a business-oriented sense, whether we speak of a bank or of a manufacturing company the applications layer of its business architecture is *not* monolithic. On the contrary, it may include a myriad of processes, attempting to observe:

- Application-to-application communications, and
- An associated message exchange control.

Able solutions are not just a matter of code conversion. The process of message normalization is fairly complex and it has to be polyvalent. Essentially the message standard we adopt says three things at the same time:

- What the message is,
- What it contains, and
- How it should be handled given its format.

A distinction can be made between *internal* and *external* message formats which support a variety of uses – instructions, reports, confirmations and so on – each with its own characteristics. In many companies, as far as the internal format or protocol is concerned, there may be different dialects according to the medium being used. These are captured differently even if they are shaped in about the same way.

Would it not be possible to abandon this format diversity and adopt a standard one, for instance like the fast packet switching we consider in a later section? The answer is a 'qualified No' because we have to care not only about international norms and what we might wish to do, but also about what our business partners *are* doing.

Even if there was no other reason, this sees to it that the formats and protocols confusion which is prevailing worldwide *cannot* change though it has created inefficiencies and incompatibilities. The careful reader will appreciate that there is an embedded contradiction between this statement and the fact that in the future a company must have formats and protocols which are:

1. *Open* for database-to-database communication,
2. *Efficient* in transaction processing,
3. *Homogeneous* in internal and external message handling,
4. *Reliable* in terms of operations, and
5. *Secure* from origin to destination.

This contradiction *might* be overcome by the fact that technology can be of major assistance both at hardware and software level. An example of the former is fault tolerance, while a software example is provided by knowledge engineering. The Sumitomo Bank, for example, has had in operation since 1987 an expert system which:

- Handles fully automatically, including error correction, some 85 per cent of all Telex messages.
- Presents structured queries for quick reply by an operator for another 10 per cent, and
- Leaves a mere 5 per cent of all incoming messages for manual handling.

Compare this to the 60–70 per cent manual handling in the majority of companies and you get an idea of improvements which are doable in a communications environment through new, more sophisticated types of software.

THE RIGHT WAY OF LOOKING AT TELECOMMUNICATIONS PROTOCOLS

The right way of looking at telecommunications protocols is as problem-solvers. They are an integral part of the system solution we are after, but they should also be able to serve in an able manner the business architecture put in place to enhance the company's impact in the market, and the efficiency of its operations.

As a matter of principle, the whole process of computation and communications needs to be automated through and through – from pre-input online at the client's site, to post-output also to be delivered online at that site. This helps to underline that:

- Protocols and formats are not just a technical problem.
- More than anything else they are a business problem, and their choice should be approached in this sense.

It serves precious little to convert from X.25 to fast packet switching when the response given to clients' requests is predominantly done by mail instead of being *online*: networked, software-supported and computer-based. Low-cost and high-quality standards suggest that:

- The solution should feature direct computer-to-database access, after authentication, and

- Fast packet-switching protocols (see the following section), not the unsecure and inefficient usual practice with 3270 and other obscenities.

Therefore, particular attention should be paid to a cost-effective and timely approach to organizational problems – and to the fact that many data processing shops are often feet-dragging. New protocols with old procedures is not an approach which commends itself. It costs money and the return on investment (ROI) is trivial.

Telecommunications solutions must permit *a quantum leap in technology*. The goal should be value-added financial services. But value-added services require a higher level of sophistication in the adopted organizational solutions – not just at technology's end.

This is essentially a cultural issue because in terms of technology the needed changes are no big mountain to climb, and they are doable over a short timetable. But they are doable provided top management is serious about seeing them happen.

Business decisions should condition the technical choices and the chosen solution should permit a smooth transition to the high-bandwidth applications of the late 1990s. If swamping high labour cost and increasing performance are the goals management are after, then a cultural change should be the primary target. Otherwise, it is fine to stay put in a stagnant technology environment.

As far as organizational studies are concerned, particularly studies done with the purpose of improving upon the current situation, it is not enough to look at what people do today in order to provide the basis for change:

- We have to project on what they will do tomorrow.
- This means we have to change the mode under which they now operate.

All this discussion is very relevant to the question of format and protocols because, as it cannot be repeated too often, the benefits from new technology can be easily wiped out by backwater organization and rusty procedures. Another domain where major attention should be taken is lifelong training. Man is a programmable engine.

Billions of dollars are spent each year on training people to use computers. Over $50 billion is reportedly expended annually on training programmes in the US alone, in appreciation of the fact that user-training is one of the critical components to successfully implementing a computers and communications system.

Sceptics ask lots of questions all revolving around the notion: 'How productive really is all this cost in time and money?' – but they fail to see that without intensive training the money spent on communications, computers and software is singularly unproductive. Therefore, the key questions should be:

- How can we train *and* motivate users to employ technology to its greatest potential?
- How do we design effective training approaches that will ultimately enhance return-on-investment?

Investments in technology will give little or no return if the human resources are not of the highest grade. Therefore companies should diligently support and assess not only their users' needs and know-how – but also that of their experts in telecommunications and software. Armed with these notions, we can now look into the notions underpinning fast packet-switching protocols.

BASIC CONCEPTS UNDERPINNING FAST PACKET SWITCHING

The concept *fast packet switching* comprises a whole array of different devices and techniques which have in common a faster throughput of voice, data and other signals. This is a field in full evolution with great impact on multimedia technologies, but different opinions exist about implementation timetables.

Fast packet switching protocols include variants around two main lines. These are known as *cell relay* and *frame relay* approaches, making feasible faster, safer and more cost-effective communication.

- With cell relay, which characterizes the asynchronous transfer mode (ATM), a fixed packet size characterizes the switching process.
- By contrast, a variable packet size is characteristic of *frame relay*, which some experts consider an intermediate technology.

For reasons which we will examine in this chapter, the best way to look at frame relay is as an approach to upgrade only what are today's medium speeds. It is not that efficient and only supports voice, but it is advisable in beefing up X.25 implementations.

By contrast, ATM permits the handling of multimedia bit streams by integrating voice, text, graphics, image and data into a common switching process. Greater efficiency is another asset, with all that this means in terms of business opportunity and competitiveness.

The problem is ATM costs. On the one hand, ATM's multimedia capability and high speed makes typical local-area networks running Ethernet or token ring look downright archaic. But price is one big hindrance.

- Installing ATM in LANs costs about $1500 per personal computer.
- This contrasts poorly with the $300 or $400 for beefed-up LAN standards like fast Ethernet.

Cost estimates are like the chicken and the egg question. *If* the market is big, *then* unit costs will drop. But typically the market will spouse a new technology and grow, *if* the unit costs significantly drop. In spite of this contradiction, I was told by cognizant people in my 1996 research that:

- By 1997/98 ATM costs will be nearly competitive to Ethernet, and
- By 1998/99 will be available an ATM Enterprise Backbone with multicast at an affordable price.

Figure 2.2 shows the fast packet-switching alternatives, their standards and the processes which they may support. Typically, fast packet switching includes many issues ranging from relaying of variable length cells up to fixed length cells, as well as standards like LAPD, Q.931, IEEE 802.6 (the IEEE 802.6 MAN standard supports the Distributed Queue Dual Bus (DQDB), and services like SMDS, B-ISDN and so on (for ISDN and B-ISDN see Chapter 4).

The groundwork on which this new technology rests lies in the fact that today there is an increasing need for high-speed wide area networks (WAN) as well as metropolitan area networks (MAN). At the same time, many local area network (LAN) installations are up for renewal. Depending on the field of application, there exist different solutions to meet requirements:

Figure 2.2 The fast packet switching concept and its implementation

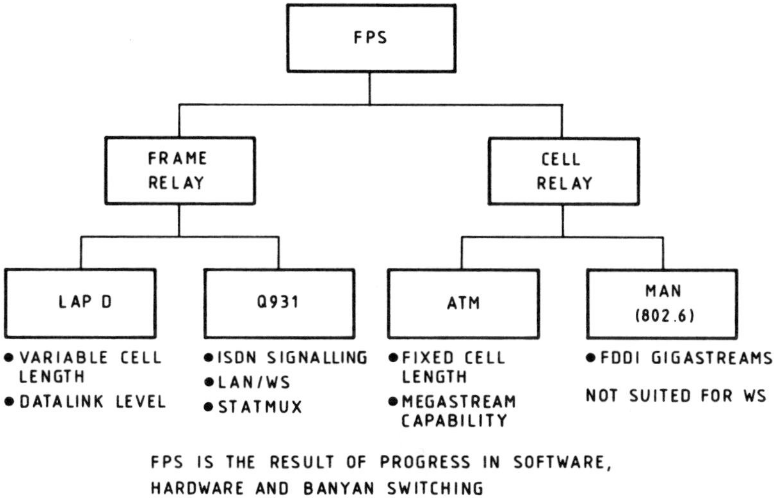

- The moving gear behind these standards is progress in hardware and software technology as well as in packet-switching principles.
- Another factor is the need for efficient solutions to face the integration of data and voice traffic as well as other challenges.

Voice and data do not have the same characteristics in what concerns response time, burstiness, traffic volumes, or other session times. But in the bottomline, it is cheaper to have a uniform transport and switching technique for both voice and data – and by extension for multimedia.

Frame relay is a first step in the right direction, because it is possible to have a combination of frame relay and circuit switching. In terms of integration, ATM is the solution for multimedia transmission because, by sharing the same channel, it opens the way of future developments.

Table 2.1 makes a comparison between ATM and frame relay features. It also contrasts the two of them to time-division multiplexing (TDM), the X.25 protocol and FDDI. The reference to OSI layers is to the Open System Interconnection model of the International Standards Organization (ISO).

Table 2.1 A comparison of communications disciplines

	Frame Relay	*ATM*	*TDM*	*X.25*	*FDDI*
Main application	data	data, voice, video	data, voice	data	data
Main benefits	low overall delay, flexible	high speed, channel-sharing	best suited for voice	good for value-added services	good solution for MAN
Main drawbacks	limited value-added functions	high cost of hardware	assigned time slots	long delay per hop	sensitive to interrupts
OSI layer	2	2/3 (ATM adaptation layer)	2	3	1, 2
Frame (or cell) length	variable	fixed: Header: 5 bytes; Info: 48 bytes		variable; 128 bytes as typical maximum	variable; 4500 bytes maximum
Standards	CCITT I.122	CCITT I.121 (B-ISDN)	CCITT G.700	CCITT X.25	X3T9.5

As it will be appreciated, none of the five alternatives has all the advantages we may wish to obtain – or all the disadvantages. Which of the 'pluses' has more weight, is largely a matter of choice in design. One of the key advantages of fast packet switches is that they boost the cost-effectiveness of digital leased lines.

Currently available ATM products allow users to implement packet switched voice and data at megastreams, or megabits per second (MBPS). There are two alternatives essentially characterizing current technology:

- 1544 MBPS over North American T1 leased lines, or
- 2048 MBPS over European E1 lines.

This compares well to conventional X.25 data packet-switching speeds of between 64 kilobits per second (KBPS) and 300 KBPS. Proponents say fast packet switches use bandwidth more efficiently than T1/E1 circuit switches. Experts also point to management advantages, such as the ability to re-route traffic more quickly in the event of line failure.

Compared with the tens of thousands of bits per second offered by conventional X.25 solutions, fast packet-switching products upgrade the more traditional approaches in a number of ways. The most important perhaps is that the re-engineered system is capable of sending traffic at a rate between 100 000 and 1 million packets per second.

Fast packets are smaller in length. But even though each packet contains less information, the far larger number of packets sent means more efficient use of the network components – from lines to switches – which has evident aftermaths of telecommunications efficiency.

Fast packet switches use a method of voice compression called *voice activity detection* (VAD). Suppliers maintain that VAD improves both efficiency and quality because silence in conversations – which stands at about 40 per cent of the total time – is not transmitted.

Vital to future solutions will be the ability to handle in an able manner multimedia databases, which we examine in Part Two. For this, we need a technology that can convert and compress voice, image and other analog signals into digital signals. This sees to it that there is interest in:

Figure 2.3 Capital spending in 1994 by European telephone companies in billions of dollars

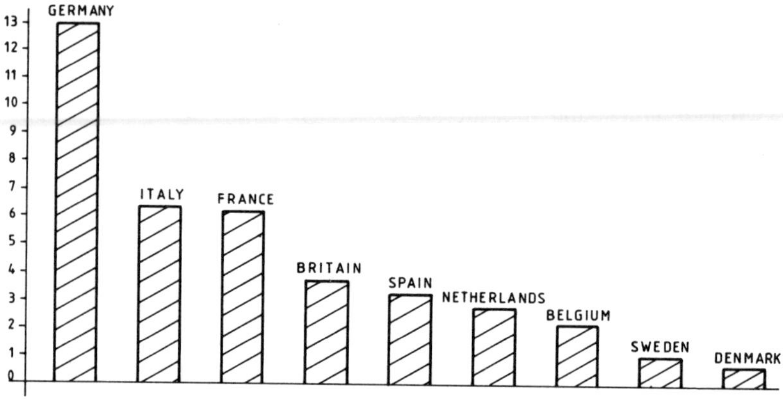

- Developing new high-level signal processing solutions, and
- Doing research in areas like algorithms for encoding multimedia signals.

This can be stated in conclusion. Fast packet technology serves as a transmission medium and switching solution with a promising future. While frame relay applies to data only, the asynchronous transfer mode permits multimedia integration and therefore seems to hold the upper ground.

NEW TECHNOLOGIES, SYSTEM SOLUTIONS AND BUSINESS OPPORTUNITIES

Chapter 1 and the first four sections of this chapter explained why it is always rewarding to look at new technologies as business opportunities rather than just a technical challenge. With fast packet switching, bandwidth is allocated more quickly and the existence of alternatives permits optimization.

How many telephone companies are taking advantage of new technologies to upgrade their communications networks? Taking the European scene as an example, Figure 2.3 answers this query by looking at the most recent capital expenditures. Notice that even if these 1994 statistics are taken on a per capita basis, Germany will remain as number one and Denmark as number nine – though the German lead in terms of telecom investments will be reduced.

- But while the lack of investments keeps a telco in backwater, money alone does not solve problems.
- The best technical solutions are those most thoroughly studied, simulated and experimented with.

Studies have shown that frame relay is suitable for traffic that travels in bursts, such as that passing between local area networks. Based on its bandwidth-as-needed approach, frame relay reduces the number of interfaces necessary to link LANs across long distances.

As we have already seen, frame relay features variable length packets (frames). The variable length makes the solution flexible and presents operational advantages, which is a point of serious debate among network designers.

With frame relay, packets are also sent more quickly than with X.25 because, among other reasons, they do not need many of the X.25 functions developed for analog circuits (frame relay is discussed further in later sections). Then, in terms of cost-effectiveness,

- Users can cut costs by employing only the bandwidth necessary for the traffic associated with a particular application.
- This is better than having to dedicate large chunks of bandwidth to specific functions, the way elder approaches work.

On the other hand, the asynchronous transfer mode protocol is designed to meet requirements of continuous media and wide bandwidth networking. Its small, fixed packet size allows hardware designers to build fast switches – though it is not optimized for any of the applications to which it addresses itself. Recent advances permit the production of switches and interfaces relatively inexpensively.

For these reasons ATM is becoming one of the preferred protocols, widely accepted as a viable standard for general network applications. In this cell-based network technology cells are 53 bytes:

- 5 bytes of header, and
- 48 bytes of payload.

Figure 2.4 Framework and scripting levels help create an architectured approach to GUI

Cell-based networks like ATM break all packets into a stream of small fixed-size cells. One difference with the classical packet-based networks of the late 1960s and the 1970s is the small size of the unit of transmission for cell-based networks. We will return to this subject in more detail in Chapter 3.

If, however, the protocol is a formalism characterized by polyvalence – as stated in an earlier section – *then* much more than the structure of header and payload has to be specified. To appreciate this reference, we should think beyond the application level of the OSI Reference Model, in an end-user-oriented sense.

A glimpse is given in Figure 2.4 which presents two layers: the framework and scripting levels which help to create an architectured approach to graphical user interfaces (GUI). *If* ATM is designed to serve multimedia streams, which is the case, *then* in terms of its implementation we should think not only of the data link level but also of GUI. As it will be appreciated, as shown in Figure 2.4, there are two sublayers:

- The one is system-oriented, and therefore macroscopic.
- The other is user-oriented, and it presents the microscopic view.

This can be said in a nutshell. The new protocols capitalize on the fact that, since the advent of VLSI chips, digital technology has played an increasingly focal role in telecommunications systems, and this role is now extended at user level. The aftermath is an expansion of communications to include the processing of images and other multimedia information.

ALTERNATIVE PROTOCOLS AND THE NECESSARY ASSISTANCE BY AGENTS

While packet-based networks typically require that the entire packet be stored on the adapter before transmission can begin, over an ATM network transmission can start with very little data. This is one of the reasons why this approach has the potential to provide low-latency communications.

Cell-based ATM standards apply today to broadband infrastructures from 34 MBPS (E3) speeds upward. For narrowband speeds, only the frame-based Data Exchange Interface (DXI) standard is applicable, but it lacks the key flexibility and performance benefits of the asynchronous transfer mode.

Only recently standards are moving forward for ATM at narrowband speeds using the fixed-length 53-byte cell format. The prevailing opinion among experts is that standards for ATM at T1/E1 speeds and below will accelerate the migration:

- From TDM-based private networking,
- To ATM-based hybrid networking.

A basic reason behind this statement is that users are looking for increased performance while controlling the costs of their communications solutions. Are there other protocol alternatives available which are worth consideration?

The answer is 'Yes!', there are. For instance, the *express transfer protocol* (XTP) is supporting TCP-type byte streams and also isochronous traffic, such as video. It has been defined to allow hardware somewhat easier implementation, but it did not manage to become very popular.

The *Internet protocol* (IP) supports the transmission of data over a collection of networks. It corresponds to layer 3 in the OSI model. Connected to the Internet protocol is the Transmission Control Protocol (TCP) which addresses the OSI protocol layer 4.

TCP is popular and it is supporting communication based on a reliable byte stream. *User Datagram Protocol* (UDP) is another Internet transport protocol for OSI layer 4, that supports best effort delivery of packets. This means no retransmitting of corrupted or lost packets – which, after all, is the characteristic of datagram.

There is a new version of IP known as IPv6, or *Internet Protocol version 6* (see Chorafas, 1996a). Most leading vendors are committed in supporting IPv6 with their products because not only does it present significant improvements over IP but also, quite importantly, it is designed to run on:

- ATM networks, and
- Low bandwidth channels – including wireless.

The *Internetwork Packet Exchange* (IPX) protocol used by NetWare takes blocks of data, normally 512 bytes per block, and packetizes them by adding the destination address and other information. These packets are sent sequentially to the destination address, with no acknowledgement required.

Fibre Channel is an optical linking protocol used mainly between workstations and high-performance computers in cases where massive data transfers must be done with the shortest possible transfer times. This protocol is optimized for bulk transfers and therefore does not apply well to multicast continuous media situations such as teleconferencing.

The *High Performance Parallel Interface* (HIPPI) was designed to meet internode data transfer needs mainly for array supercomputers. The hardware required for HIPPI is inexpensive, but does not scale beyond LAN implementations due to the limited length of its signal lines as specified in the protocol. It supports a point-to-point channel at 800 MBPS or 1.6 GBPS.

HIPPI is in evolution. Proposed enhancements may improve this distance to fit MAN requirements. The problem is that the bandwidth has to be multiplexed and switched in parallel over the WAN, which is generally not done, though plans are being made to interface HIPPI with ATM multiplexers.

Tn and *FTn* are other alternatives. T1 and T3 trunks and fractions of their bandwidths (FT1 and FT3) are a common medium for carrying voice and data in a WAN. The lines may be dedicated copper or optical fibre, or sub-bandwidths in very-high-speed links.

There is also FDDI, considered by many to be a protocol whose time has come and gone. In this sense, FDDI has something in common with another technical anachronism: ISDN. Therefore they will both be treated in Chapter 4 – a deliberate choice.

Within the ISO/OSI architecture we should also pay attention to the *media access control* (MAC) sublayer. It is defined by IEEE as the lower portion of the data link layer in the OSI Reference Model. The MAC sublayer is concerned with media access issues, such as whether token passing or contention will be used.

Though brief, these references respond to the query about protocols – yet the answer is not complete. As it has been stated on repetitive occasions in this chapter, the solutions we are after should not be limited to the choice of a line discipline; though we absolutely need an efficient protocol for multimedia transfer.

Just the same, however, we need increasingly sophisticated software to help us manage the network. In the previous section we saw a brief example with the GUI. At this point I would like to add another example which addresses network management requirements, and most specifically the use of knowledge engineering in diagnostics.

As a matter of principle, we should always be keen to learn from the best applications available today, rapidly proceeding with the implementation of what works well. For instance, the General Telephone and Electronics (GTE) Laboratories have made a first class application known as CALIDA (California Intelligent Database Assistant), which provides seamless access to heterogeneous databases. This is one of the functions an intelligent network should support at user sites.

Figure 2.5 gives a bird's eye view of the design underpinning CALIDA. Its functionality rests on four pillars which, among them, constitute a good frame of reference for applications involving interactive knowledge engineering artefacts, or *agents*.

- In the case of the GTE application, the solution targetted seamless access to incompatible databases.
- In other implementations, agents can handle very diverse functions such as multimedia support, diagnostics and end-user assistance.

Agents are a fairly recent development which became popular at a rapid pace. They are also an excellent example of the strategy followed by tier-1 companies in order to gain advantage from the building blocks of network intelligence. The able choice of protocols is inseparable from this subject – but it is not the only issue to which attention needs to be paid.

Figure 2.5 Conceptual approach to the design of CALIDA

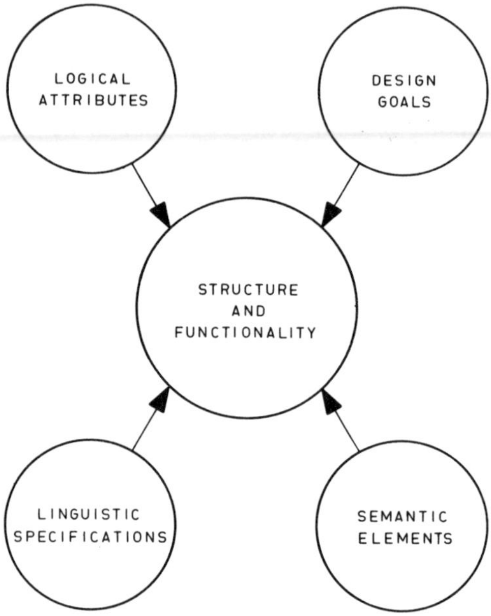

COST FACTORS, BANDWIDTH SHARING AND NETWORK MANAGEMENT WITH ATM

Chapter 3 is dedicated to the technical features of the asynchronous transfer mode. By contrast, the goal of this chapter is to bring under perspective business opportunities connected to the effort of refining a system to make ever higher-grade communications possible. Most evidently, this has changed the requirements connected to the choice and implementation of protocols.

The problem is that many companies not only do not appreciate these requirements, as should have been the case, but have not even heard of the business benefits of ATM. What they mainly see is an engineering-driven technology which is still too expensive to apply.

- Current prices seem to constitute the single most important issue to potential ATM users, as stated in an earlier section.
- Next in line comes the query how to integrate ATM with an existing communications infrastructure.

Companies knowledgeable about the use of technology will look to at least a 5-year timeframe. In this sense, they will buy ATM if they are convinced it will cut costs and improve customer service over this period of time. Therefore, education on direct benefits to be derived – not just the technical aspects – is key to ATM's business success.

Analysts who have considered this subject stress that there is an impressive number of issues underlying ATM's implementation. These range from the business case of services, to technical issues such as LAN emulation, network management, interconnect challenges and applications drivers.

Bandwidth sharing and statistical multiplexing are two of the issues which have been studied extensively over the past few years – because they both constitute important technological solutions. This work lead to several analytical approaches and simulation models that:

- Compute the equivalent bandwidth of a session, and
- Do so when an application is multiplexed with a number of other sessions.

But the complexity of interconnections and allocation algorithms increases with the evolution of the implementation environment. While the homogeneous traffic case is relatively simple, heterogeneous traffic can pose difficult problems. One of the alternatives, the *equivalent bandwidth* computation, is too time-consuming to be carried out in realtime at call setup.

The problem is that contrary to what mainframers and centralized network management adherents have been saying, since bandwidth preallocation techniques are imperfect, they also expose the user to congestion. They do not respond to user traffic parameters stipulated at call-setup time.

Management of a large variety of multiple sources with different traffic profiles and requirements is a tough nut to crack. Let us keep in mind that in conventional packet networks, bandwidth sharing was not a difficult problem because the constraints were different.

X.25 networks lacked strict delay and bandwidth guarantees, as well as powerful control mechanisms. By contrast, in ATM networks bandwidth sharing becomes much more difficult to enforce, and this for two reasons.

1. The reactive flow and congestion procedures are too expensive to support on high-speed links, and they are still inefficient.
2. Multiple classes of traffic, which characterize multimedia, do not have homogeneous requirements.

At this point in time, for bandwidth management and congestion control reasons the ATM solution rests on the premise of a homogeneous network where all traffic is transformed into a uniform 53-byte packet or cell. This permits the network to carry a wide variety of different traffic types suitable to heterogeneous user data streams.

Three strategies address the bandwidth management problems. One is *peak rate* allocation, which permits the user to specify the maximum rate at which cells are to be sent to the network. However, if traffic exceeds a specified rate, cells are simply discarded. This poses its own problems and makes poor use of bandwidth in the case of bursty traffic.

The second strategy is known as *minimum throughput* allocation. The user specifies the throughput that is needed when the network is congested and the network guarantees the specified throughput. The third strategy is *bursty traffic* specification, allowing the user to specify:

- Peak cell rate,
- Average cell rate, and
- Maximum burst size.

Network management software uses these parameters to configure the network in an effort to assure the specified allocation can be met. The main problem with this approach is that it may take a long time to compute when a new data stream needs to be multiplexed, since the procedure is computationally intensive.

Confronted with problems caused by heterogeneous traffic, ATM service providers are now looking at *virtual subnetworking*. This permits subdivision of the traffic into more homogeneous groups aiming at sharing bandwidth with other customers having unknown traffic characteristics, by constructing a *virtual private network* (VPN) on top of the backbone ATM.

The enabling protocol is the *virtual path* (VP) mechanism. It permits establishment of a virtual connection between any two points – with possible allocation of peak bandwidth. The user can multiplex several *virtual circuits* (VC) within the VP which he regards as a privately-owned facility.

As this discussion suggests, the protocol is only one component of a system solution. Many other issues enter a rigorous study, some of them being in full evolution. Yet their solution is vital to a successful implementation. ATM, like frame relay that we study in the next two sections, is no ideal solution – but it increasingly looks the better alternative.

TAKING A CLOSER LOOK AT FRAME RELAY

The best way to look at a frame stream is as a physical circuit configured to transport frame-relay packets in a dependable and uninterupted manner. For higher reliability, link integrity must be supported by monitoring *permanent virtual circuits* (PVCs).

Technically, frame relay is well suited to LAN applications where bursty, low to average bandwidth requirements are typical. For LAN/MAN/WAN internetworking applications, frame relay provides a dynamic bandwidth allocation which serves individual LAN users with required high-speed WAN connections.

- Economic advantage is gained by sharing the high-speed WAN connections amongst many LAN users.
- This must be done in a way seamless to the user and effective in applications terms.

Solutions which serve this goal are worth careful study in an implementation sense. On the basis of the above two points, some telecommunications vendors, like Northern Telecom, today promote frame-relay technology – but as already stated in an earlier section on fast packet switching, the majority opinion in the telecommunications industry is that frame relay is an intermediate technology not a long-term solution.

As we saw in an earlier section, frame relay is one possible implementation of fast packet-switching technology, using variable length frames. It serves the goal of having a fast interface between end-user equipment and the network.

- The frame structure of frame relay is ISO Layer 2, as described in the CCITT Rec. Q.931.
- This protocol permits packet switching to be done at reasonably high speed, albeit for data streams only.

Some vendors promote frame relay for LAN-interconnection (see also in Chapter 4 the discussion on FDDI). Above the data link layer, frame relay is protocol transparent, but its implementation requires frame-relay access devices (FRADs), for customer premises, to hook into a public frame-relay service or to link sites to a private frame network.

A FRAD is practically needed for every public frame relay service connection. This equipment comes in several forms. Some can be bridges and multiprotocol routers. Others may be dedicated asynchronous devices or multiplexers, as well as hybrids of these designs.

FRADs transmit data through digital circuits to switches, from local area networks, terminals or individual computers. With the appropriate support, the frame-relay protocol, designed to work over error-free digital connections, offers greater throughput but lacks the error correction and recovery of the more robust X.25, which was designed to work over unreliable circuits.

One of the claims made by its proponents, is that frame relay is more efficient than X.25 for short bursts of high-bandwidth data – because it does without the error correction and sophisticated addressing of the older protocol. But a result achieved at the expense of dependability is not necessarily welcome:

- On the positive side, each switch on the network has less processing to do and so can pass on packets more quickly.
- On the negative side, if errors do occur, the entire packet has to be retransmitted.
- Also, if there is no sequencing messages can arrive in the wrong order – and the receiving end must rearrange them, which is a datagram procedure.

Because of the current heterogeneity of equipment, and the problems faced by some of the receiving terminals, I personally do not like this approach. Therefore, I foresee that X.25 will continue to serve low-bandwidth business applications. Understandably, however, bandwidth-hungry customers are looking at alternatives to X.25 to boost throughput without significantly increasing costs.

My thesis is that while it is clear a more efficient protocol than X.25 is needed, users will be reluctant to move away from what they view as a reliable packet-switching service. The evidence to this hypothesis is that some vendors tend to integrate frame-relay offerings with their existing X.25 products.

Stated in a different way, frame relay's most commonly cited technical vice is the absence of a reliable, fully implemented scheme for *flow control* – which has business implications. If flow control were a purely technical question it would have long since been relegated to obscure standards committees and vendor forums; but it is a business proposition as well.

Necessarily, a debate on protocols integrates other technical questions. In particular, the way of handling higher bit-error rates on many trunk infrastructures, which contain a great deal of copper and microwave bandwidth (see also the silly business of using twisted-pair for fast packet switching, in Chapter 4). Such amalgamation of technologies may create problems for frame relay's minimalist approach to error detection and correction.

- Even a small increase in error frames can cause rapid congestion, as end-system transport protocols retransmit the missing frames.
- The result may be a brutal throttling down of network throughput, which would cancel out the low-delay argument – one of frame relay's chief virtues.

Technical worries such as these have led some frame-relay critics (and X.25 supporters) to suggest that affordable VLSI semiconductor processing power now makes possible X.25 switches that can fill up 2 MBPS trunks with very low transit delay. That may be true but the argument misses the point that X.25 is, by now, an obsolete protocol and it is only data-oriented.

FRAME RELAY AND X.25 MARKET OPPORTUNITIES

Readers with experience in system design will appreciate that every protocol has its strengths and weaknesses. Frame-relay studies done on an experimental basis tend to suggest this protocol is best suited for 4- to 10-node networks. Based on this and other findings it can be argued that two different frame-relay market segments may be shaping up:

1. The LAN interconnect market, which is small in volume but highly competitive.

This market is based on service offerings from established international X.25 and value-added network players. It also includes, to a certain extent, connectivity to the more traditional network of international leased lines.

Without frame relay, network interconnection from multiple locations requires multiple physical links for each end-point device. For instance, in a 10-node network, each node needs nine links plus the dedicated bandwidth to be fully interconnected, further increasing installation and operations costs while, correspondingly, reliability decreases.

2. The domestic data-oriented leased line markets, served principally by monopolistic or oligopolistic telcos.

With the introduction of the frame-relay interface, frames are multiplexed to and from the backbone via a single high-speed link. In a network sense, this preserves the original mesh topology and reduces port and line costs. Because bandwidth is dynamically allocated, there are also savings in operating costs.

Due to their high prices, leased lines in the international market are already in the process of being displaced by packet switching. In domestic markets, the split between X.25 and leased lines depends on the historical attitudes of local telecoms. The fact however remains that, so far, frame-relay traffic:

- Has moved only across corporate backbone networks or over the public network.
- Somehow, it has not interfaced between private and public networks.

The appeal of frame relay over X.25 will be greater if a documented and factual assurance about the availability of reliable, error-free transmission – not just high bandwidth – can be given to the user organization.

There are also other subjects to be brought to the reader's attention. For reasons explained in the preceding paragraphs, a spate of frame-relay products and services raises concern about the interoperability of the network equipment outfitted with interfaces built to a nascent standard. At issue is whether users will be able to easily interconnect diverse networking devices, such as routers and multiplexers,

- To each other, and
- To public frame-relay services.

To date, the small number of users that have added frame relay to their corporate networks have done so on a limited basis, working closely with their vendors to asssure interoperability. Interoperability of various vendors' frame-relay equipment is key not only to the sales of the equipment itself, but also to the success of the standard. In principle,

- Before determining whether their equipment is interoperable, vendors should make sure their products conform to frame-relay specifications.
- The problem is that such specifications are not yet well settled, making it difficult to establish the appropriate conformance tests.

Interoperability is so much more important as vendors are selling frame relay like bandwidth-on-demand. What is being promised for delivery is not so much a guaranteed minimum, but a guaranteed maximum average over time. Users, however, will be well-advised to ask both for the minimum and the maximum values to be guaranteed in contractual terms.

Here again, the lack of universal standards is regretable because both vendors and users need to conduct conformance tests throughout all phases of product development and implementation. This means from the definition of requirements and applications design, to operations and maintenance.

In other terms, not only must vendors use conformance tests to ensure that their protocols and communications links are in line with frame-relay specifications, but also user organizations must be in a position to verify the vendors' claims. They also need to keep well in mind that conformance does not guarantee interoperability in spite of the fact that:

- If a vendor's product conforms to frame-relay specifications,
- Then it is more likely it will interoperate with other compliant networking equipment.

In conclusion, among the benefits of frame relay are the low overall delay due to high-speed access lines and the flexible bandwidth allocation. This aspect is important as, in connection with the variable length of data packets, it is possible to cost-optimize telecommunications solutions.

- Frame relay offers the user the possibility to send a large amount of data at a high speed.
- But one of the requirements is that capacity is available to exploit the advantages of fixed time slots.

Another basic need is dependability assurance, not just a vague reference that 'traffic will travel over clear digital lines'. Provided dependability is guaranteed, frame relay has major advantages over X.25, including lower delay at switching nodes and protocol transparency above the link layer. A 'plus' of frame relay is also the fact that another protocol can be encapsulated in a frame – which is not possible with cell relay.

But we have also spoken of drawbacks: voice applications are not supported and no standard collision control mechanism has been developed. Problems could arise when several LANs are connected to a frame-relay network. Most importantly, there is no error correction mechanism – and this is a significant 'minus' for many applications using voice-grade lines in a long-haul network.

Finally, no standard tool for monitoring the network load is currently available. And there are other drawbacks which lead vendors to integrate frame relay and X.25 features into their equipments. Users will be well-advised to be very careful with their mission-critical solutions – at least until frame-relay technology develops the necessary dependability in methods and tools.

3 Appreciating the Implementation of Asynchronous Transfer Mode (ATM)

INTRODUCTION

To its promoters, the asynchronous transfer mode (ATM), also known as cell switching, is the best available on-premise internetworking technology. Its high bandwidth, uniform switches are workhorses for multimedia. They can transfer graphics, audio, video and text from application to application at much higher speeds than now available.

- The asynchronous aspect of this technique lies in the fact that a single channel does not send the information in synchronized time slots.
- By all likelihood, the asynchronous transfer mode will be the more extensive implementation of fast packet switching.

ATM started life in Cambridge University. It then became the basis of massive development efforts aimed at building the next generation of networks. In recent years, it has been actively promoted by major telephone equipment companies and many Silicon Valley start-ups.

There is hardly a firm in the networking game that does not have either an ATM product or an ATM strategy. Deutsche Telekom, for instance, plans to have a good part of its operations on ATM by 1997. These are national plans. Worldwide, many technical problems still have to be addressed – the queuing of messages being one of them. Finite buffering does not seem to smooth the queue.

The ATM Forum claims most technical standards are complete, but that does not mean equipments made by different companies will work together without problems. In a lot of cases the standards are not subject to supervision and compliance, and ATM devices made by different vendors may not work together.

In terms of strategic developments, many datacomms vendors have set their sights on the design and marketing of network superhubs that bring together today's smart hub with router and wide area transmission features, all on top of an ATM switch:

- ATM is seen as the technology that helps to integrate the computer and communications industries.
- Eventually multimedia communications between workstations and servers will be handled in that way (see also the discussion in Part Two on multimedia servers).

Applications which have so far been completed using ATM technology suggest that there are many benefits from this approach. After all, what are our alternatives in terms of new communications technologies? Chapter 2 has answered this query and, as the reader has appreciated, there are no other star performers.

UNDERSTANDING THE NOTION OF AN ATM CELL

As it has been explained in Chapter 2, the ATM protocol is based on a *cell*, or fixed-size virtual circuit-oriented packet which characterizes the system's switching methodology. The reason for choosing a fixed-size cell was to assure that switching and multiplexing functions could be carried out quickly. Quite evidently, fixed cells have limitations. In this sense, frame relay is more flexible.

By using a standard cell size ATM can switch data via hardware. This is more efficient and less expensive than software methods, the way X.25 and frame relay work. Other things equal, hardware-based switching is faster:

- ATM speeds are scalable and can exceed 2.5 GBPS over fibre.
- ATM also provides features necessary for successful multimedia applications.

The asynchronous transfer mode has the ability to define different traffic types, each delivering a different *Quality of Service* (QOS).

Figure 3.1 Structure of an ATM cell: header and payload

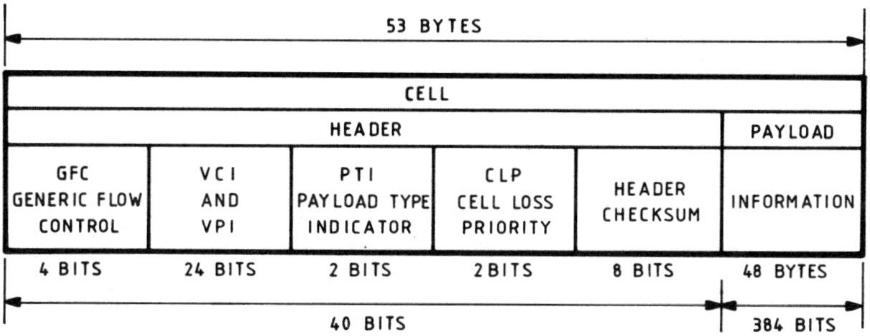

One of its modes of operation, the Constant Bit Rate (CBR) supplies a fixed-bandwidth virtual circuit. We talk of CBR in the next section. Its line discipline addresses the special handling needs of delay-sensitive multimedia applications containing realtime video and voice.

The size of the ATM cell has itself been a compromise. As shown in Figure 3.1, such cell consists of five bytes of header information and a 48-byte information field – a total of 53 bytes. The header field contains control information for the cell. Specifically,

- Flow control,
- Virtual circuit identification,
- Payload type indicator (PTI),
- Cell loss priority (CLP), and
- Header checksum (the header is protected by an 8-bit cyclic redundancy checksum – which, however, does not cover the payload).

The protocol breaks all traffic into these 53-byte cells which in essence has been a middle-road approach addressing different applications. Telephone companies wanted a small cell to reduce delay for voice packets. Data communications experts preferred a big cell to minimize the amount of segmentation and reassembly that had to be carried out.

- For data communications, a cell length of about 1500 bytes would probably have been a good value.

- This, however, is a hypothesis rather than a statement because much depends on the application itself and its bit stream.

Transmission of the bits within a cell is synchronous. But single channels do not have reserved time slots as in the case of time-division multiplexing. Therefore, the grand design is an asynchronous transfer of information.

Two fields of the header are of particular interest. On each incoming link of an ATM switch there is a *virtual channel identifier* (VCI) and a *virtual path identifier* (VPI). These are used to determine which cells belong to any given connection. The routing subsystem employs VPI and VCI to determine the:

- Path(s) and
- Channel(s)

a cell will travel. Cells belonging to a connection do not always have to appear after one another in the data stream. They are statistically multiplexed, with the amount of allocated bandwidth determined by the traffic requirement of the connection.

Information carried by VCI and VPI uniquely determines the new virtual identification safeguards to be placed in the cell header. Also the outgoing link over which to transmit the cell. This specification is most vital because:

- The concept of a virtual path permits a set of ATM connections to be manipulated as one unique channel.
- Two hosts can use a virtual path to multiplex many individual application streams.

All of the fields of the ATM cell header are important in LAN, MAN and WAN interconnection: generic flow control, facilities for virtual path and virtual channel, payload type indicator, and most importantly the cell-loss priority field.

While VCI and VPI are used for channel identification and simplification of multiplexing, payload identification helps to distinguish between user cells and control cells, permitting a separation of user and control information. For its part, cell-loss priority indicates whether a cell may be discarded during periods of network congestion.

A significant reference in terms of interconnection is that the func-

Figure 3.2 Traffic concentration in an asynchronous transfer mode solution

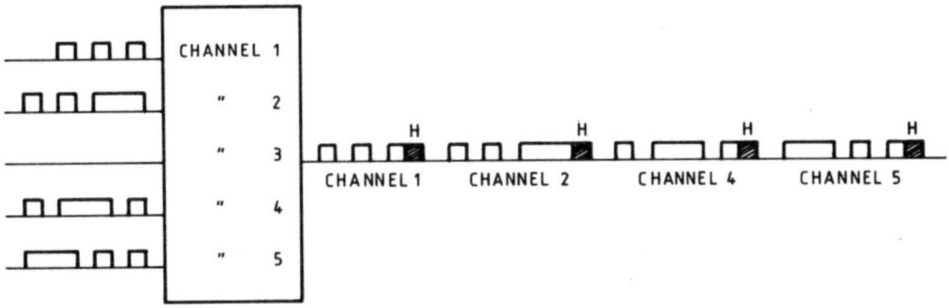

tions performed by the ATM layer are carried out in hardware at gigastream transmission rates. This is a tier-1 competitive advantage since it permits greater speed and lower cost.

This relatively brief technical review helps to explain what is meant by the statement that ATM works through a *cell-relay* approach. It also assists in identifying a main difference between cell relay and frame relay – which lies in the cell length convention: Cell-relay packets have a fixed length, but an important feature of ATM is that a cell is not dedicated to a channel.

- When a channel carries no data, there is no reserved time slot for that channel as is usually the case with TDM.
- Each ATM cell carries information of only one channel, as can be seen in Figure 3.2.

Because the length of an ATM cell is a compromise, experts disagree on whether or not the fixed cell length in ATM is a drawback. In my judgment this is not necessarily a severe constraint. A fixed cell length is a way of doing the switching with pure hardware, nearly as fast as physically possible – and this process has a price.

The give-and-take in design is important because ATM grew out of the world of switching networks with virtual connections between nodes, rather than the world of LANs. In the latter case, nodes share a medium and transmit by passing tokens or sensing for packet collisions. We will further examine this issue when, in a later section, we talk about usage of the ATM protocol in connection to LAN, MAN and WAN.

ATM FORUM, AVAILABLE BIT RATE (ABR) AND OTHER METRICS

Recently, the ATM Forum defined a new class of ATM traffic which complements so far available metrics. The *Available Bit Rate* (ABR) service is aimed to provide realtime access to spare network bandwidth. ABR uses rate-based, closed-loop feedback mechanisms. Other ATM service disciplines currently available include (in alphabetic order):

- *Constant Bit Rate* (CBR) which provides a constant bit rate pipe for applications such as voice, circuit emulation or continuous bit rate video according to JPEG/MPEG. JPEG stands for Joint Photographic Expert Group; MPEG, for Moving Picture Expert Group. Both have established standards.
- *Variable Bit Rate–Realtime* (VBR–RT) which transports variable information, providing bandwidth for bursty traffic, like LAN interconnection. As long as rates do not exceed a specified average, it provides tight bounds on delays.
- *Variable Bit Rate–Non Realtime* (VBR–NRT). It transports variable rate information to support applications like transaction processing.
- *Unspecified Bit Rate* (UBR) service. It targets best effort delivery, but advances no Quality of Service guarantees while using available bandwidth.

Table 3.1 compares ABR with CBR, VBR and UBR. Notice that the available bit rate discipline is the only one to provide for feedback-controlled sharing of spare network bandwidth. It is supporting protocols such as TCP/IP, IPX/SPX, and APPN and particularly addresses implementation environments which do not have fixed or predictable bandwidth requirements.

With CBR and VBR the user must declare parameters at connection set-up. But there are many applications for which it is impossible to accurately predict in advance these parameters. The facility ABR provides is that what the user needs to do is to state the maximum bandwidth possibly needed and (optionally) a minimum usable bandwidth which the network should guarantee in the event of congestion.

It is left to network feedback to dynamically adapt the rate offered to the user based on available bandwidth. This permits ATM service providers to maximize utilization of their network with flexible

Table 3.1 Classes of service and their characteristics

	Bandwidth Guarantee	*Throughput Guarantee*	*Delay Variation*	*Congestion Feedback*
ABR	Yes	Yes	No	Yes
CBR	Yes	Yes	Yes	No
VBR	Yes	Yes	Yes	No
UBR	No	No	No	No

approaches which can be adaptable on the basis of bandwidth availability.

UBR services also try to fill wasted transmission slots, but this line discipline has no feedback and makes no QOS guarantees. Also, during network congestion, there may be a high cell loss. Therefore, either service providers overprovision their network leading to lower overall performance or users have to do with low actual throughput, considering the retransmission taking place due to cell discard.

For their part, the realtime and non-realtime alternatives of VBR traffic consume variable amounts of real resources, while bandwidth must be reserved to meet the quality of service requirements. As a result, network utilization is characterized by wasted transmission slots – which reduces throughput.

All this changes with ABR services because closed-loop feedbacks prevent congestion, and adapt traffic to the available bandwidth in realtime. Feedback control and buffering provide better bit rate service and make feasible to implement:

- Explicit Forward Congestion Indication (EFCI),
- Explicit Rate Markup (ERM)
- Segmented Virtual Service/Virtual Destination (VS/VD), and
- Hop-by-Hop VS/VD disciplines.

EFCI switching represents the minimum required to comply with rate-based congestion management standards. ERM features an active means of participating in congestion notification, marking resource management cells.

The Segmented Virtual Source/Virtual Destination facility breaks the feedback loop into segments, providing a congestion control firewall. This insulates the network from misbehaving users. Hop-by-hop VS/VD breaks the feedback loop into individual segments, one per pair of switches; it also offers a tight control, but it is complex and expensive.

Tuning is important. Buffer allocation and the size of ATM switch buffers impacts on performance. Typically WAN switches require larger buffers than LAN switches. A VS/VD switch per virtual circuit buffers at access interfaces permitting traffic management to be directed to specific connections.

Optimization can be provided by First-In, First-Out (FIFO) buffering schemes. In this case different users share the same buffer space. This approach, however, cannot assure a dynamic fair share allocation of bandwidth to all users. Therefore, it lowers the guarantees for quality of service.

The notion of quality of service advanced by the ATM Forum should attract considerable attention. Vendors are addressing problems like synchronization by assigning a quality of service to applications. QOS identifies an application as:

- Time-sensitive, and
- Requiring priority.

Typically applications of lower priority are less time-sensitive. There are however major queries in a design sense. For instance: Where is the best place to handle QOS? If it is the network, at which level: routers, hubs, switches? The operating system? Or some combination of hardware and software?

EVERY IMPLEMENTATION INVOLVES CHOICES IN DESIGN

With the implementation of ATM, voice and other traffic will be transmitted in a packet format. In a TDM approach, available bandwidth is divided up into time slots of a fixed size (see also the discussion of multiplexing TDM and other protocols to ATM later in this chapter). A better implementation allows the allocation of bandwidth on demand. But the more sophisticated the possibilities are,

- The more complex become the algorithms which have to be used to divide up the bandwidth optimally, and

- The more important that a chosen solution should lead to short delays within the multiplexers.

Many of the proposals on the internal architecture to characterize an ATM switch are still at a research-and-evaluation stage. Various objectives, such as blocking, routing, performance and VLSI implementation, have motivated past work. More recent issues address concerns which are important to the commercial ATM market, such as the general cost of ownership as well as incremental:

- Reliability (see Chorafas, 1960),
- Scalability,
- Deployment, and
- Efficient bandwidth usage.

A key benefit to be obtained from a system with dedicated channels is the minimization of buffering. But there is the problem of *bitslips* as well as of *silent time*. In the case of classical telephony, about 50 per cent of the time there is silence. This percentage slightly depends on the language. In an ATM solution silent periods are not wasted but used to support other channels – which is a major advantage.

Capitalizing on its features, the asynchronous transfer mode supports high-speed integrated services – which makes ATM a good solution for backbone technology.

But as the Introduction underlined, many technical problems still await solution, the smoothing of message queues in ATM networks being one of them. For this purpose the Weibull distribution is recently being used, which:

- Was originally employed with reliability studies in guided missiles, and
- Most recently is used in connection to derivatives risk-management.

With ATM, routing is not needed on a per-packet basis, as required with present-day approaches, but is still necessary on a per-connection basis. The same algorithms and functions supported today by a router still apply in the ATM world except in a connection-oriented configuration.

Advantage is taken of the fact that although ATM operates within the physical layer, it can also perform virtual circuit addressing and routing. Optimization is feasible because the asynchronous transfer mode protocol supports:

- Both constant and variable bit rate,
- Connectionless and connection-oriented framing, and
- Necessary multiplexing services.

These are among the reasons why the asynchronous transfer mode is a primary-candidate technology for bringing multimedia solutions to the client–server paradigm. Advantage is taken of the fact that, as we saw in an earlier section, cell routing is based on a two-level addressing structure, the:

- Virtual path, and
- Virtual channel indicators.

The first identifies the physical path, whose functionality is then exploited by a set of virtual channels. The generic flow control field defined at the user network interface can implement crude congestion control. Congestion is one of the crucial competitive features of a protocol.

Competition to ATM comes from the fact that frame-relay technology – which we studied in Chapter 2 – has a possibility to combine its data-only traffic with voice traffic based on circuit switching. The sense of this combination is that the same backbone carries frame-relay data and non-frame-relay voice signals. Such a solution, however, can be complex as it is necessary to install frame-relay interfaces into the network switches.

Finally, the commercialization of ATM, just like any other network technology, mandates a set of requirements to be considered in an architectural proposal which encompasses both multimedia and single-media. These include:

- Coexistence with existing technologies,
- Upward scalability of ATM-based networks,
- Efficient ways to extend the switching bandwidth,
- Dependability and reliability factors, and
- Costs – from ownership to operations and maintenance.

It is fairly sure that these requirements will play a more crucial role in future research on ATM switch architectures. The speed with which ATM services develop is partly a question of technological maturity and partly one of market pressure. The latter translates into competitive positioning among equipment vendors and service providers.

Finally, in spite of the fact that the ATM arena is starting to become congested in terms of vendors, I strongly believe that demand for ATM implementations will not take off until software that effectively exploits its features is available. At the same time, sophisticated software requires a good-sized market. It is a chicken-and-egg problem.

For the time being, ATM suppliers find it difficult to deliver on one of their central promises: that ATM networks will effectively handle voice along with text, video and data. There are technical reasons behind this statement:

- Data can be sent in irregular time chunks,
- But voice requires a steady, unimpeded flow.

While the theory of how to do the necessary parallelism is known, actually delivering good-quality voice at an acceptable price has been a subject which, so far, seems to have overwhelmed suppliers. Eventually, some vendors will come up with valid solutions – and will capture the market's heart.

LOOKING AT ATM AS A BACKBONE TECHNOLOGY

The ATM switches on the market today embody attributes that meet current demands, while at the same time they continue evolving to meet future requirements. As Figure 3.3 suggests, the backbone solution is, most likely, the first major phase of ATM implementation. This will be followed by a bifurcation towards:

- Desktop interconnection on one side, and
- Intercompany networks on the other.

There are of course a number of technical challenges to be addressed in all three applications areas. By using small cells for data

Figure 3.3 Milestones in the development of ATM

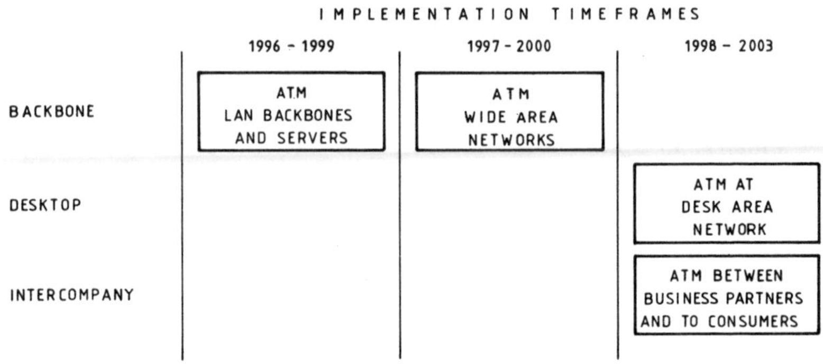

transfers, ATM needed to support multiple forms of data. This differs from other high-speed networking technologies in that it is the first that encompasses both:

- Local area network (LAN), and
- Wide area network (WAN) perspectives.

Much has been modelled on the idea of a worldwide telecommunications network. Many experts suggest that, as a backbone technology, ATM will work best in campus environments – as a high-speed LAN backbone replacement – but can also find a good market in wide area networking solutions, as a leased-line replacement by carriers.

For instance, if the user organization requires four links, carriers are now starting to offer the alternative of a single physical connection into the site with four logical connections at digital signalling level-3 (DS-3). This will support speeds of 44.76 megabits per second, and it is preferred to the more expensive option of four leased, dedicated lines.

At the same time, as a backbone solution, ATM does not feature the queuing and processing delays in an X.25 packet-switching network. The network is designed for higher throughput and low delays due to:

- High-speed trunking,
- Hardware-based switching, and
- A simplified protocol.

Channel sharing is important particularly with multimedia where data, text, voice and video signals are transmitted over the same channel in a direct database-to-database. We will talk more of this in Part Two in connection to multimedia servers.

Backbone solutions also capitalize on the adaptation-layer interfaces between ATM and higher layers of the ISO/OSI model. Most functions have end-to-end significance, and the use of many among them is optional. Within this perspective, the International Consultative Committee for Telephony and Telegraphy (CCITT) has defined four classes of ATM services. In a nutshell these are:

- *Class 1*: A constant bit rate application, such as pulse code modulation telephony.
- *Class 2*: A variable bit rate non-data application, like compressed video.
- *Class 3*: A connection-oriented data application.
- *Class 4*: A connectionless data application.

Class 1 addresses the case where a time relation exists between the source and the destination. The bit rate is constant, and the service is connect-oriented – as for instance in a voice channel. A time relation exists between the source and the destination also with Class 2, but the bit rate is variable with the connect-oriented service also addressing image transmission.

By contrast, with Class 3 no time relation exists between the source and the destination, and the bit rate is variable. The service includes connect-oriented file transfers. Similarly, with Class 4 no time relation exists between the source and the destination and the bit rate is variable. The difference is that this service is connectionless, as in the cases of:

- Local area network interconnection, and
- Electronic mail applications.

Within the perspective established by these four classes, the ATM adaptation layer plays an important role in the internetworking of different structures and services. By interconnecting local, metropolitan and wide area networks it aims to support applications with diverse traffic characteristics as well as multiple data streams with guaranteed delay bounds.

The described serviceability is enhanced by the fact that each ATM adaptation layer protocol has two sublayers. The lower sublayer, or segmentation and reassembly protocol, breaks the data into 48-byte cells and collects them into packets. The upper convergence sublayer prepares the user data for the lower sublayer.

Both vendors and users try to capitalize on the efficiency and flexibility. In August 1995, for example, IBM unveiled its Switched Virtual Networking strategy for migrating router- and SNA-based networks to switched ATM networks. The mainframer has put together an approach to the handling of multiprotocol environments that seems to address the needs of both LAN and SNA networks.

In terms of user organizations, one of the best examples is SuperJANET, the advanced academic communications network that deployed ATM switches in a nation-wide structure. Other examples employing the same technology by GDC are: PictureTel which provides realtime communications; MCI with its HyperStream ATM; MFS Datanet offering an end-to-end international ATM service – as well as France Telecom and Telecom Finland, among network services providers.

But as networking vendors rush to the broadband multimedia communications based on ATM, significant strategic differences are emerging. To understand the background reasons, we must appreciate that asynchronous transfer mode was designed to let public networks carry every conceivable form of digital traffic. Now it is being embraced by the data-communications industry, which is looking at ATM as the solution to all kinds of networking problems.

1. Some say the fibre-based, 45 MBPS-plus ATM approach is ideal for high-speed image-based workstation applications.
2. Others think it would be a perfect all-purpose solution if it ran over cheap, copper twisted-pair wiring.
3. Still others are of the opinion that the application for ATM is that of a backbone, campus technology; an improved version of data-only FDDI.

We will examine points 2 and 3 in Chapter 4, but let us conclude this section with the following thoughts. Currently, among the main benefits of ATM is speed of transmission and protocol structure – also its future potential. It is assumed that speeds beyond the 140 MBPS on broadband channel will be cost-effective, while the switching speed will possibly range from 100 000 to 1 000 000 packets per second.

THE APPEAL ATM PRESENTS TO MULTIMEDIA SOLUTIONS

The asynchronous transfer mode has been described as the technology that allows total flexibility to be achieved in broadband multi-service, multimedia networks. As such it has received a great deal of attention, particularly in connection to integrated services and universal connectivity enabling high-speed time-critical applications solutions to reach the workstation.

While, as Chapter 2 and previous sections of this chapter demonstrated, the basic ATM technology presents interesting features, it has also been said that there are still a great many questions to be answered. For instance,

- How will multimedia applications effectively use ATM?
- Which will be the best software support approaches possible?
- What kind of specific implementations will be enabled by this technology?
- Which other breakthroughs might be possible within a multimedia environment?

Starting with the fundamentals, industry increasingly requires networks with high bandwidth and low latency, to provide realtime transport capabilities necessary for future multimedia applications incorporating voice, text, data, graphics, image and animation. As we have seen, ATM seems to answer this need.

Within the evolving multimedia implementation perspective, a more technical issue is the fact that the asynchronous transfer mode is making major strides towards becoming the mainstream interface technology – not only for database-to-database communications but also for multimedia computing. However:

Figure 3.4 Three layers of an asynchronous transfer method protocol and the applications environment

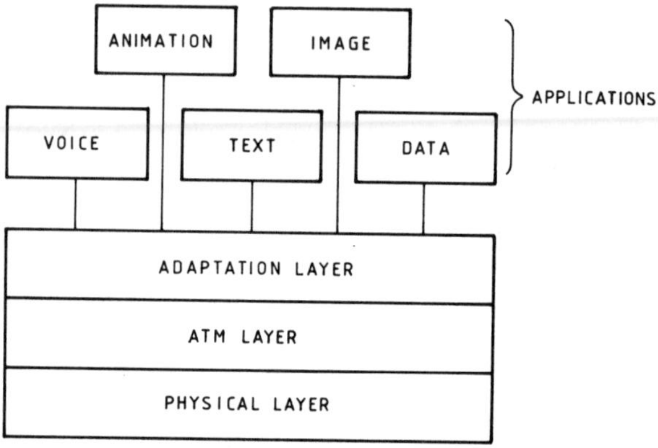

- Costs will have to come down significantly for this to materialize, and
- The right client orientation would be not the consumer but corporations.

In a corporate setting, advantage can be taken of the fact that unlike shared-media topologies, such as Ethernet or FDDI, ATM is a switch-based cell relay that connects individual nodes over a dedicated bandwidth. By statistically multiplexing identically-sized cells over one line, ATM can handle a mix of data, text, voice and video, including realtime video-conferencing.

- Because such technical characteristics are appealing, the synchronous transfer mode is given a boost.
- But to satisfy broadband requirements in realtime, new network architectures must be designed and implemented.

ATM's physical and logical aspects assist in this direction or in rearchitecturing. The protocol reference model is based on standards developed by the International Telecommunications Union (ITU) and is divided into three levels, as shown in Figure 3.4:

- ATM Adaptation Layer (AAL),
- ATM layer, and
- Physical layer.

The physical layer defines a transport method for ATM cells between two ATM entities. It also features a medium-dependent sublayer responsible for the correct transmission and reception of bits on the physical support.

An interesting feature is the transmission convergence sublayer focusing on the mapping of ATM cells to the transmission system being used. We will look at the role of the logical layers when, in the next section, we talk of ATM in terms of LAN, MAN and WAN integration. Hence, it is important to keep this sublayer reference in mind.

Because multimedia pose challenges which are not necessarily present with single-media solutions, ATM is increasingly seen as a general protocol to link desktop computers across the corporate landscape for high-bandwidth applications. However, to do this it needs to be implemented in LANs and internetworking devices – at an affordable cost.

- It should also be possible to build internetworking ATM products from different vendors.
- But in terms of an open vendor policy standards are still incomplete, and this is indeed a pity.

Another issue making ATM particularly interesting to multimedia solutions is that, unlike a circuit-switched network, an asynchronous transfer mode neither tracks nor switches the messages running through the network fabric. Rather, the messages tell themselves where to go.

- The frames of the data have their headers read on the fly.
- This is not classical packet switching but, rather, distributed data communications.

Important also is the fact that ATM networks do not need to be controlled by centralized resources, because the packets can find their own way to their destination. In effect, they ask each node they meet whether it has the address to which they have been sent.

In conclusion, ATM networks are attractive because of their flexibility and the potential of network control savings due to their self-routing algorithms – which will be most valuable with multimedia. Yet, not to be forgotten is the fact that the most advanced distributed networks need sophisticated software to keep tabs on what is happening and to interpret the records.

Let us also keep in mind that, despite its appeal, ATM has possible drawbacks. Data cells can potentially be lost and never retransmitted in congested environments. In a manner similar to frame relay, there is no guaranteed retransmission function – which can eventually pose problems with some multimedia applications.

There also exist other negatives as, for instance, the fact that crucial technical requirements in areas such as congestion control, signalling and management are still missing. And, for the time being at least, carriers show little appetite for providing public ATM services at affordable prices.

Indeed, today the price to implement ATM is one of its main drawbacks. Therefore, prior to a wider mission-critical implementation, an investigation has to be made regarding the overall cost of the new system. Multimedia solutions are important – but not at any cost.

ATM, VIDEOCONFERENCING, VOICE INTEGRATION AND PNNI

Because of its ability to significantly improve productivity in groupwork, a fast growing application with ATM is desktop videoconferencing. For this reason, even if they are adequate in the near-term, campus, metropolitan area, and wide area backbones are migrating to connection-oriented environments with ATM. Such transport mechanism:

- Permits scalability in speeds,
- Provides flexibility to support multiple virtual connections, and
- Makes feasible service guarantees with personalized bandwidth on demand.

ATM-based solutions support a virtual environment able to accommodate frequent adds, moves and changes, desktop-to-desktop.

This is one of the reasons why ATM is emerging as networking technology of choice for videoconferencing and other multimedia applications – supporting high bandwidth communications with low latency. It can statistically multiplex video streams over a single line, rapidly switching available bandwidth from one stream and one address to another.

Asynchronous transfer mode solutions deliver the capability to allocate bandwidth for a specific quality of service required for high bandwidth applications. The ATM protocol can also execute usage parameter controls to assure that each video connection receives the required level of performance from the network.

An ATM to the desktop standard is intended to lead to a terminal device supporting voice with a direct ATM interface. This can permit making calls over the public voice network. Technical questions revolve around the signalling plane and the voice plane. The current answer is 64 KBPS PCM, but its implementation details are difficult to define. The goal is to permit a desktop ATM device:

- To make voice calls to another ATM devices, and
- To do so across a public voice network.

The best way to look at computer-telephony integration (CTI) is under the aspect of multimedia applications. Not only do they represent the most advanced features but also propel us into new ways of working. But before we can effectively realize the potential of CTI, we have to address the challenge of effectively delivering multimedia applications over LANs, MANs and WANs, keeping in mind that:

- Multimedia traffic is time-sensitive, and
- Different digital streams, especially voice, must arrive at their destination at the right time.

If the audio portion of a videoconference is out of synchronization with the video components, the result is low quality of service. One of the problems connected to effective multimedia applications is bandwidth. Another is protocols. For instance.

- IP does not allocate a specific path or amount of bandwidth to a particular session.

- Therefore, the resulting delay can vary unpredictably – which is unacceptable with realtime applications.

Hence, the Resource Reservation Protocol (RSVP) proposition developed by the Internet Engineering Task Force (IETF). RSVP runs on top of IP to provide receiver-initiated setup of resource reservations on behalf of an application data stream.

When an application requests a specific quality of service for its data stream, RSVP delivers the request to each router, maintaining router and host states to support the requested level of service. The RSVP program applies at each router an admission control to determine if it can supply the requested QOS.

- If yes, the RSVP program in each router passes incoming data packets to a packet classifier,
- The classifier queues the packets in a packet scheduler which allocates resources for transmission on the link.

RSVP works well with IP version 6 (IPv6), allowing users to set up end-to-end connection for a given time. Video and voice get special handling. In this manner multimedia applications can identify packets that belong to time-sensitive traffic flows.

More streamlined solutions can be obtained in private settings, where *private* implies use within ATM networks featuring private addressing, as contrasted to public addressing. A *Private Network-to-Network Interface* (PNNI) is a protocol for use between private ATM switches and among groups of ATM switches.

- A key feature of PNNI is a hierarchical structure which scales well for larger ATM networks.
- Routing is sensitive to quality of service requirements, and PNNI has the ability to automatically configure itself in networks.

The drawback is that the PNNI is complex. Therefore, to address the short-term need for developing and testing multivendor networks, the ATM Forum adopted an interim protocol. The *Interim Inter switch Signalling Protocol* (IISP) can be deployed quickly. However, IISP requires manual configuration of route tables, has scalability limits, and will not interoperate with PNNI implementations. All this suggests that there are significant choices to be made in network design.

ATM IN AN ENVIRONMENT OF LAN, MAN AND WAN INTEGRATION

ATM cells may be carried in an externally framed synchronous transmission or in a cell-based asynchronous transmission. As we saw in the last section, when we spoke of the physical layer, it is the business of this layer to encode and decode the digital strings into suitable electrical and optical waveforms for:

- Transmission, and
- Reception

on the communication medium being used. The adopted systems solution must also provide cell delineation functions, performance monitoring, and payload rate matching of the different transport formats used at this layer. ATM has these features over and above the header error check-sum which has been missing from older protocols like X.25.

The function of the ATM layer identified in Figure 3.4 is to carry all the different classes of services supported within a 53-byte cell. This is where transparent transfer of fixed-size 53-byte cells or ATM-layer *service data units* (SDU), between communicating upper-layer entities, is defined.

The main mission of the ATM layer is to perform switching activities responsible for cell relaying between ATM-layer entities. This leads to cell multiplexing of individual connections into composite flows of cells, but also:

- Cell demultiplexing for composite flows into individual connections,
- Cell-rate decoupling or unassigned cell insertion and deletion,
- Priority processing and scheduling of cells, with cell-loss marking and reduction.

Other functions include cell-rate pacing and peak-rate enforcement, explicit forward congestion marking and indication, cell payload-type marking and differentiation, as well as generic flow-control access.

In a LAN, MAN and WAN setting, the mission of the ATM Adaptation Layer is to provide a link between the services required by higher network levels and the generic ATM cells. This level consists of a sublayer that assures cell segmentation and reassembly to interface to the ATM layer, as well as a more service-specific convergence function to interface to the bearer services being carried.

Another important function for a modern protocol is to perform a multicast operation efficiently, as many collaborative applications will require frequent use of this kind of function. Furthermore, a good protocol must be interoperable with existing local, metropolitan, and long-haul networks as well as being integratable with existing standards.

This emphasis on local, metropolitan and wide area networking is important because one of the implementation domains of ATM networks is backbone operations – as discussed earlier in this chapter – and another is distributed computing. In the latter environment,

- An application program can be partitioned and distributed across a network of computers and software modules, and
- Solutions must capitalize on the fact that the virtual path concept of ATM allows virtual connections to be set up between the various processing nodes.

Within each computational pattern, a way should be found that best suits the application(s) requirements. This helps to reduce the time needed to route messages between nodes, and permits a constant pipelined data flow between applications running on different nodes.

A good example of how an ATM solution integrates into a LAN environment can be provided by ATM to IP links. Sophisticated methods for assuring that asynchronous transfer mode and *Internet protocol* devices communicate are starting to emerge. The *Internet Engineering Task Force* (IETF) has released a specification for running IP atop ATM.

Unlike earlier methods, this approach will let IP networks tap ATM's advanced features, such as mixing realtime videoconferences with file transfers. One of the better approaches is LAN emulation, where Ethernet and token-ring LAN traffic rides over an ATM network.

- The ATM network emulates the LAN environment's media access control (MAC) layer.
- But though LAN emulation does the job of mixing today's LANs and ATM, it does not directly address the IP/ATM connection issue.

The main attraction with LAN emulation is that it preserves existing LAN applications and does not entail any complex network reconfiguration. It is a simple way to get applications into an ATM environment. A more complex but comprehensive approach is the overlay method, where IP traffic runs over ATM. On the downside, however, this:

- Does not provide the IP environment with ATM's advanced features, and
- Does not support IP-based video traffic.

A more sophisticated approach is the *Multiprotocol Over ATM* (MPOA), where existing networks like IP backbones get the real benefits of ATM such as mixing voice, data and video traffic. MPOA is coming up with a way for multiple IP subnetworks within an ATM network to send traffic directly to ATM devices rather than through routers.

- A device on an IP legacy network would go to its nearest router to transmit traffic to the ATM network.
- Experts, however, think that MPOA is at least four to five years away.

Still another ATM integration approach is the peer method: all devices on a network, including IP-based, would communicate as peers. But this solution requires outfitting ATM switches with multiprotocol routing knowledge. IP traffic going across the ATM network would have to be mapped into an ATM address structure. No one expects this method to take off anytime soon. But the concept of routing in the ATM switch is alive in both the MPOA and peer methods.

In conclusion, an efficient new protocol must allow for two facts: first high-bandwidth implementations will generate a heterogeneous pattern of network traffic. Second, we have to accommodate connectivity to older networks.

Neither can be served through an easy solution. Hardware and software diversity is very difficult to efficiently support with protocols. Also, current networks do not provide the transport facilities necessary to support increasingly complex user applications. Hence, our interest in IP/ATM and other connectivity issue is at this time by default rather than the result of rigorous testing.

SYNCHRONOUS AND ASYNCHRONOUS TRANSFER MODES FOR MULTIPLEXING REASONS

Typically, the development of component technology determines both the multiplexing scheme in networks and the relative growth in switching and transmission technology (see also the discussion in Chapter 4 on synchronous and asynchronous line disciplines). It also influences the balance between switching and transmission being employed. In the post-World War II years, the main multiplexing techniques have been:

- Space Division Multiplexing (SDM),
- Frequency Division Multiplexing (FDM),
- Time Division Multiplexing (TDM).

In regard to network interconnection, each has its advantages and limitations. This is particularly true as networks provide heterogeneous services and therefore require efficient approaches for multiplexing and switching.

A brief background would help. Manual switching has been the earliest solution in telephony. By the end of the 19th century, the invention of mechanical switching by Strowger replaced manual operations. Strowger exchanges and their variations typically lasted till the end of World War II, *at which time* they were replaced by electronic crossbar switches.

The earlier switching solutions followed *space division* principles. A telephone conversation exclusively used a path in the switching network for its entire duration. Means were also required for efficient control of switching solutions as well as for diverse features to characterize telephone services.

Space division offers multiplexing capabilities but the approach classically taken is not that flexible. The development and use of reliable analog amplifiers, as well as microwave and coaxial transmission, required more efficient protocols which were provided by *frequency division* multiplexing (see also Chorafas, 1991).

Once multiplexed, analog voice signals could be amplified and transmitted at lower cost over wide ranges. But the cost of multiplexing grew significantly with the number of signals multiplexed.

- This made wider band transmissions suitable only for long-distance and high-volume routes.

- The development of FDM tended to consolidate the long-distance network into fewer routes with higher capacity.

However, as telephone companies started working with digital technology, different signals were multiplexed in the time domain to form a single stream for transmission on a communication channel.

Time division multiplexing sees to it that we no longer have to dedicate a single transmission channel for a single communication connection. Digital communication provides the possibility of making a better choice – which is not just a technical problem but relates to the economics of telecommunications services.

- An optimal switching algorithm rests on identifying at an early stage sets of functionality to be supported at affordable cost.
- Technico-economic considerations impact all stages of development, including basic infrastructure, equipment and applications.

For the economics of telecommunications to work in a sufficiently efficient manner, solutions must be found which answer intercommunication requirements. One of the key advantages of TDM over FDM is that the multiplexing cost per signal decreases as the multiplexing speed increases.

With lowered cost supported by computers, TDM became more cost-effective than FDM. TDM can be seen as more than multiplexing and transmission, as it can also be used in *Time Division Switching* (TDS). This makes the distinction between transmission and switching less obvious in local area network architectures, where communication terminals are distributed along a shared transmission system.

Further, the deployment of fibre optics and gigastream (GBPS) rates of transmission, as well as the technology for coherent lightwave detection, promoted *Wavelength Division Multiplexing* (WDM). For long-distance transmission, optical TDM transmission became more reliable and cost-effective, per bit of information, than analog microwave FDM transmission.

- Optical transmission technology has become the vehicle of choice for network integration.
- Its huge and reliable transmission capacity can accommodate an impressive number of services.

Another crucial advantage is that optical transmission technology can, at the same time, make feasible further developments. The more technology advances, however, the more problems need to be solved.

In the absence of a common reference such as terminal synchronization or realtime scheduling, simultaneous transmission generates interference among the attached terminals. One communication discipline which addresses these conditions is *Code Division Multiplexing* (CDM).

- CDM bandwidth may appear in time, frequency or space.

CDM distinguishes itself from TDM, FDM, or SDM by allowing the possibility of simultaneous transmission by different terminals at the same time, frequency, or space.

- The CDM protocol achieves reliable communication by redundantly encoding the information of each terminal.

In this way the transmitted signal can be uniquely decoded despite interference from simultaneous transmission of other terminals.

In the absence of a scheduler, one multiplexing technique employs random access algorithms which assume that conflicts recursively resolve their contention through random retransmission. This is one of the models used for deploying public end-to-end fibre networks.

- But for optical networking, synchronization can be difficult.
- Therefore, it has to receive a very substantial amount of attention on an international scale.

One of the current approaches assumes a common time reference, or frame reference, among the terminals. This mode of multiplexing is commonly known as the *Synchronous Transfer Mode* (STM). Another approach assumes no frame of reference among the terminals, hence the *Asynchronous Transfer Mode* (ATM) which we have already examined.

WHAT CARRIERS AND USERS THINK OF ATM APPLICATIONS

As 1994 came to an end, Sprint invited a host of American high-technology companies to develop applications over its 240-kilometer broadband network in California. These applications included desktop videoconferencing, transmission of dense graphics and video images, and distance learning.

The emphasis on multimedia is understandable (see earlier in this chapter). Sprint suggests that new applications projects should allow users to access the most up-to-date graphics and related software online. Other implementations must permit companies at different sites to work online.

ATM seems to be a good candidate in connection to these solutions – provided that the objectives we are after are clear, and whatever tactical adjustments may be necessary we know where we are going.

- Some telcos think that ATM approaches might be instrumental in developing software for films and TV commercials, as well as for publishing.
- This sounds reasonable, but as the following short case study suggests, it is not a self-evident truth – neither is it a method free of technical adjustments.

Caledonian Publishing, publisher of the world's oldest English-language daily newspaper, has conducted extensive trials of the next-generation technology focusing on ATM. Like most other major publishers, this company needs broadband networks to:

- Transfer the huge data files involved in electronic newspaper production,
- Implement its business plans to carry live video feeds to its clients, and
- Be able to perform its future functions through high-technology while abiding to tight deadlines.

In its environment, Caledonian Publishing has found out that average page sizes are between 20 and 50 megabytes, and hundreds run around the network in the run-up to deadline. Storage for production systems alone is more than 80 gigabytes.

Presently, Caledonian is using a Fibre Distributed Data Interface (FDDI) backbone and switched Ethernet to desktops, which include more than 200 Apple, Sun and other workstations. ATM has been proven in year-long on-site trials on workstations and in the backbone.

Still, Caledonian is not ready to move forward with commercial implementation of ATM because the input/output of most workstations is not yet sufficiently fast to fully utilize the 10 MBPS the new system offers. This position has been taken in spite of the interest in image processing present in senior-management plans.

Although Caledonian thinks ATM would yield significant improvements at the downstream end of the process, particularly in the production of deadline-sensitive laser output, it is hesitant to switch to a new technology for such mission-critical applications. There is also the question of costs. As we have seen in the preceding sections,

- While asynchronous transfer mode switches handle voice, video and data simultaneously, and
- They are working high speed in a dependable way,

the problem is that they are still expensive. One of the reasons for higher costs is that these switches are based on the crossbar design, with every junction needing its own computer chip. As of mid-1995, however, MMC Networks (of Santa Clara, CA) has come up with a design that it says could cut the cost of ATM switches within five years. The idea is to use high-speed memory instead of the grid.

- All input lines can dump their bits into this pool, and
- The computer will assign a specific digital address for the bit stream.

In the past, designers have thought of this solution but kept away from it because it requires a very wide pathway to shuttle bits in and out of the memory. But MMC seems to have discovered a way to get around that problem through a new bus design. We shall see.

In terms of cost-effectiveness, a good benchmark for memory-based ATM switching is a comparison to *switched Ethernet.* Switched Ethernet dedicates 10 MBPS of bandwidth to each user, and this solution is currently less expensive than ATM. It is also widely seen as a relatively painless means to upgrade congested Ethernet LANs.

But many specialists suggest that the transfer of the 25 MBPS channel to the desktop, which will be necessary with multimedia, would help ATM vendors compete with switched Ethernet. The ATM desktop interface will answer bandwidth requirements with multimedia, while switched Ethernet cannot handle multimedia applications in a scalable way.

Much can be done by capitalizing on the fact that ATM provides full duplex channels with isochronous capability. The challenges are both technical and economic, and with time they will be solved. The problem is that to bring costs down we need a mass market – and to develop a mass market we must have low costs.

USING ATM IN PAYMENT SYSTEMS

It has not yet been widely recognized that one of the best implementation areas for ATM technology is *payment systems*. The globalization of finance sees to it that within a reasonably short timeframe payment and settlement procedures will need to undergo a restructuring revolution largely based on communicating databases.

- This will require much more than multimedia approaches, broadband channels and the software supports available today.
- It will call for high security, and with it *knowledge* artefacts which can answer protection requirements in an able manner.

Autonomous knowledge artefacts can be used not only for payments clearing services but also in connection to security management. *Agents* help to establish and enforce security policy and services, controlling both physical access and guarding logical access to critical resources including programs, files, terminals and abstract assets such as the use of commands.

Some financial institutions are currently developing agents to protect against virus attacks. Others are working on knowledge artefacts able to exploit audit trails. Still others aim to provide a single system image of security connected to payments and settlements procedures – including one stop logon, logoff and password-change facilities.

The other leg of the effort of promoting secure ATM solutions in local, metropolitan and long-haul networks is found in the fact that some Wall Street firms have already decided to move to ATM-switched LAN. The firm Donaldson, Lufkin & Jenrette (DLJ) is an example.

One of the basic factors behind the DLJ move to ATM was the decision in April 1994, by the Federal Reserve Bank of New York, to levy a fee for trades not settled immediately. Known in financial services industry as *daylight overdraft* (intraday) these charges were designed to change the situation where traders could deal with impunity within a two-week settlement period without having to put up any funds.

A direct consequence of daylight overdrafts is that brokers have to rethink and redesign their networks and workstation software. The new directives by the Federal Reserve essentially mean that a financial institution has to get an increasing number of transactions done in a *shorter* amount of time.

- The use of ATM can allow trades to be completed faster.
- This procedure is minimizing the risk of daylight overdraft.

In the case of DLJ, transaction times have been reduced by 50 per cent since the ATM LAN went live in August 1994. The new switch takes one-fifth of a second to complete a trade, while the previous FDDI LAN took half a second.

In other terms, ATM's potential is better positioned to handle the company's increasing volume of business. The FDDI LAN began to deteriorate as the network was struggling to keep pace with the increased load. The ATM LAN also allows applications other than trading to run during trading hours. DLJ is now considering whether to integrate voice and video on its ATM LAN.

Other financial institutions in Wall Street are moving to ATM LANs because of competitive pressures. By assuring faster trading times, the added bandwidth of ATM is offering greater efficiency in handling clients and cost savings to the firm. No wonder that brokers are implementing ATM faster than other commercial organizations.

But not everything has yet been settled in this domain. Working together, major user organizations and vendors have to define the scope of work needed to bring appropriate levels of security to asynchronous transfer mode products and services. Some highlights have been defined by different studies and include:

- Authenticating end-points and switches,
- Providing data confidentiality and integrity between end-points and switches,
- Protecting ATM security signaling, and
- Creating and distributing an ATM security key.

The latter is an encryption mechanism decipherable only by the intended parties of the data exchange. Each feature will likely be incorporated at the ATM layer and the ATM adaption layer of the technology's 53-bytes cell structure. Both issues were examined in an earlier section in connection to the role of ATM in an environment of LAN, MAN and WAN integration. But it has also been said that there is need for a golden horde of new systems studies.

IMPLEMENTING ATM IN INSURANCE AND AIR TRANSPORT

Kobe-based Tokio Marine is Japan's largest non-life insurance company. Till 1994 the company used an aging SNA network. Data centres in Tokyo and Osaka served 72 IBM 8100 computers in 26 large and medium-sized branch offices via dedicated 19.2 KBPS LUO links. More than 3300 SNA 3270-type terminals in smaller branches were connected to the IBM 8100s via 9.6 KBPS LU2 links.

With such old technology, user response times were poor – often as long as 10 seconds per transaction – while anyway the network lacked the resiliency to recover quickly from disaster. Besides this, IBM announced it would discontinue support for the 8100s and attached terminals.

- With growth capacity another constraint, Tokio Marine decided to migrate to a new, open-systems solution, based on 10 frame-relay switches.
- The company also implemented Encina/DCE for transaction processing and other applications in Unix-based servers at the offices.

Routers provide protocol conversion for connectivity to local TCP/IP-based LANs. The formerly naive 3270 user-terminals have been replaced with workstations.

The new network cut user response times in half, from 10 seconds to less than 5 seconds. Disaster recovery improved dramatically. After the January 1995 Kobe earthquake, agents used redundant frame-relay links to access a backup server in Tokyo. This allowed them to meet customer needs in time of crisis. Tokio Marine is now implementing client – server solutions and, in parallel, productivity-enhancing end-user computing. The changeover takes place while experiencing a migration to ATM networking.

Seen from a broader perspective, fast packet-switching services are becoming quite popular in Japan, where the Network Information Service (NIS), a major frame-relay provider, predicts 400 per cent growth for the next year in the number of subscriber connections for its international frame-relay service. Other major providers of international frame-relay services in Japan are KDD, International Telecommunication Japan (ITJ), Japan Cable & Wireless, and Société Internationale Télécommunications Aéronautiques (SITA).

Air France presents a good example of conversion to fast packet switching in the air transport industry. The company is building an ATM-based wide area network with a high-speed backbone. This new network will:

- Replace the existing TDM-based solution,
- Carry all Air France's corporate multimedia applications, and
- Also be used to offer value-added services to other companies.

The networks being replaced by ATM include voice and data WAN's, LAN's and video conferencing set-ups, plus a variety of legacy traffic running on SNA and X.25. The ATM backbone is due to be fully installed by mid-1996 and will extend the existing network capability of Air France with:

- 100 nodes in France, and
- International connections.

Altogether, 25 countries in Europe, North America, North Africa and the Indian Ocean are connected by the change.

Cognizant executives, however, were to suggest that a magnitude of complexity separates the management of an ATM network from that of a TDM network, which features simpler point-to-point connections; or a router-based network which does not guarantee different classes of service.

For instance, resource planning which means anticipating future equipment and software requirements by extrapolating on the growth pattern, can be very complex in an ATM network. Existing modelling tools used in TDM networks and router-based solutions are not adequate to support ATM's guaranteed levels of service.

Hence, there is a need for sophisticated modellers able to provide online network design capabilities, allowing a projection of the best network topology based on current and future equipment, as well as on each connection's demands. There is also need for optimizers to analyze how expected growth or a new service will impact the network, including failures and service level – as well as installation and operations costs.

There is also a need for effective cost allocation. An ATM network's high-traffic volumes call for a sophisticated billing system, based on actual usage. This means collecting statistics at a high frequency and being able to report accurate usage information down to the virtual connection level.

This can be stated in conclusion. Introducing new high-speed telecommunications services into an increasingly competitive marketplace is a very challenging process. Before launching a new telecom service we must research the target markets' size and requirements, analyze tariffing, identify what we will offer, determine our competitive advantages and develop a strong business case for each offering – to substantiate our position and thus generate subscriber interest and sales.

4 Synchronous Digital Hierarchy, SONET and Other Protocols for Telecommunications

INTRODUCTION

The *Synchronous Digital Hierarcy* (SDH) is a set of standards from the International Telephone Union (ITU) for high-speed, optical fibre-based transmission – starting at 155 MBPS, with a current upper limit of 13 GBPS. The open interfaces of SDH aim to foster equipment interoperability, freeing operators from proprietary end-to-end systems.

Some telecom experts look at SDH as the European version of the methodology which has been employed in America with SONET, which is examined later in this chapter. Both solutions capitalize on packet streams whereby a synchronous communications line is simulated. To a significant extent SONET and SDH are very similar with the exception of one byte in the frame of the packets. But also, so far, both provide a poor security mechanism, which is a significant weakness.

Chapter 3 has explained why the asynchronous transfer mode is a fundamentally different method for switching. Fairly similar arguments can be made about the synchronous digital hierarchy mode, in terms of the signals transport:

- Another similarity is that ATM and SDH are software-based solutions allowing, among other things, automatic reconfiguration of the facility.
- To appreciate SDH's impact, we should keep in mind that modern, state-of-the-art transmission systems operate at rates nearly three orders of magnitude faster than the switching nodes connecting them.

Furthermore, both ATM and SDH are software-intense. Software-centred approaches permit systems to be upgraded more easily and controlled more effectively. They also represent a new challenge for vendors and users, who must not only develop but also constantly upgrade their programming routines.

- The software-based integrated switching and transmission allow operators to provide hybrid virtual services.
- Telcos can offer what appears to be a dedicated channel with all the bandwidth the user needs, when he needs it.

Therefore, SDH and ATM should be seen as part of the same evolution, each depending on the other to operate most efficiently. An underlying condition is that the network between any two pieces of equipment is of high quality – which means with very low bit-error rates (BER). This requires both good transmission links *and* effective switching solutions.

THE CASE OF SYNCHRONOUS LINE DISCIPLINES

Many experts in telecommunications are currently advancing the hypothesis that successive generations of SDH will generate a shakeout among transmission equipment vendors, similar to the one we have seen in digital switches. But it will also bring a synergy between switching and transmission.

Capitalizing upon this synergy implies a cultural change. For long decades, switching and transmission people operated in a way that had little to do with each other. But the dual effect of SDH and ATM stimulates a convergence of interests and of activities.

The synchronous digital hierarchy seems to offer the higher-quality and most versatile transmission links available today. There is an important intrinsic difference between the quality of a circuit going through an asynchronous network, and a circuit going through a synchronous network.

- Seen under different viewpoints, including availability, it turns out that synchronous solutions tend to perform more effectively. But there are also constraints.
- To better understand what SDH can offer, it is wise to review

some basic differences between *synchronous* and *asynchronous* line disciplines (see also Chapter 3 on synchronous and asynchronous transfer modes for multiplexing).

In asynchronous data transmission, speed is established when the call is set up, but the line is active only when bits are transmitted. When information is not being transferred the connection is still in place, but the line is effectively dead.

Asynchronous transmission is character-oriented: it starts and stops with one character, actuated through its *start* and *stop* bits. There is no clocking or timing involved, which means there are often idle times on the transmission line. Since there is no stream of data each character sent is preceded by a flag bit to alert the receiving end. The completion of the single-character transmission is punctuated with another flag bit.

- This is the model used with the earliest type of communications protocol known as *start–stop*,
- Start–stop protocols are still employed today with low band lines, because of the very low investment cost they require.

By contrast, the communications line used by a synchronous device transfers data packets according to a clocking signal. In other words, the system works with a clock. Packets are sent when the buffer fills or when an interval timer expires – but transmission is timed.

Synchronous communications are done in packets, a packet being composed of header and trailer, data field, as well as start and end flags. The clock for synchronous transmission may be originated internally from a modem or externally, as in AT&T's DDS service where one master clock times all transmissions in the country. The modems are kept in synchronism transmitting flag sequences over the link during idle times.

- Network operators have gained efficiencies by upgrading their synchronous communication transmission rather than their switching systems,
- This was reasonable, since economies of scale are easier to achieve in high-capacity transmission than in low-capacity switches.

Figure 4.1 Enterprise-wide communications systems enriched with artificial intelligence

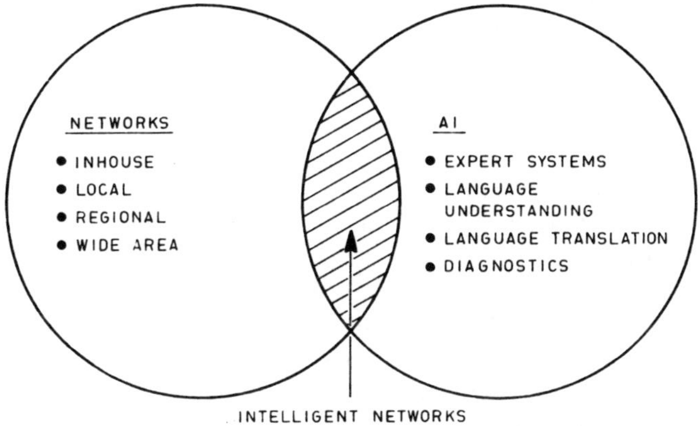

However, the introduction of ATM has created an incentive for public network operators to move further by investing in intelligent transmission systems. Their goal is to effectively distribute ATM switching power around the network supported through intelligent artefacts.

Contrary to the definition originally advanced in the mid-1980s by Bellcore and adopted by CCITT (see also Chorafas and Steinmann, 1990), a network cannot be characterized as being *intelligent* unless it is supported by an impressive array of knowledge-engineering artefacts. This requirement goes well beyond the known dependence of networks on software.

In essence, both in terms of hardware and of software a large number of upgrading systems and procedures must be put into place. The topics of local interest range from lines, trunks and switches to diagnostics. At the beginning of Chapter 2 we saw an example with CALIDA, by General Telephone and Electronics.

Figure 4.1 builds upon this reference defining an intelligent network as the intersection of the more classical communications characteristics in installations which may be made in-house, in local area, region or metropolitan, and wide area – with artificial intelligence (AI) constructs which may be expert systems, language understanding algorithms, language translation procedures or a broad array of intelligent diagnostic routines. Some years ago many of these advanced facilities were unknowns, but quite often in the depths of the unknown one finds business opportunities.

THE BUSINESS PERSPECTIVES OF SYNCHRONOUS DIGITAL HIERARCHY (SDH)

Clear-eyed telco executives will be the first to comment that old plant is a drag on the change of technology. Telcos lose a great deal by sticking to twisted-pair lines, which have been depreciated many times over but are still in service in massive numbers. By so doing they leave the door open to new competitiors, as two different examples in this section demonstrate.

The near-sighted approach by telecommunications companies to the technical and investment problem is that the bulk of transmission networks belonging to telephone operators still consist of the aging generation of Plesiochronous Digital Hierarchy (PDH) equipment. This operates at speeds up to 565 MBPS, but no more.

- PDH lacks the management header embedded in an SDH channel, providing addressing, timing bits and other data.
- The SDH structure makes the transmission channel more intelligent, simplifying and accelerating restoration in case of failure.

Deutsche Telekom is the most aggressive operator in the world in deploying SDH. It has already spent nearly DM 1 billion ($700 million on SDH equipment, and is devoting even more to developing its own SDH network management software. Synchronous technology now accounts for about 25 per cent of its overall transmission facility.

Granted, Deutsche Telekom had an incentive for doing so. Capacity requirements and the need to replace obsolete equipment in eastern Germany have been the drivers behind its SDH push. But all over the world telcos are faced with similar cases of plain old telephone service (POTS) and its obsolescence. With deregulation, short of new investments the market lies open to competition.

For instance, one of the companies adopting SDH and planning a new backbone network is Hermes Europe Railtel. Modelled along the line of WilTel, a US long-distance company that grew rapidly in the 1980s by selling fibre optic capacity alongside its parent company's oil and gas pipelines, Hermes will be a cross-border network along Europe's railway tracks.

- The Hermes fibre network is based on synchronous digital hierarchy technology, accommodating speeds of 2.5 GBPS.
- It is expected to be completed by 1999, interconnecting a significant number of major European cities.

Hermes will be a wide area communications competitor, but will not build any local distribution networks. Its strategy is to be active well before its whole network is completed. The long-haul venture is expected to be operational in six or seven European countries by 1997, selling exclusively to:

- Different operator consortia,
- Public network operators,
- Value-added network service providers, and
- Mobile phone operators.

The stated business goal is to help in meeting Europe's demand for cross-border communication services, in a way complementing the expansion activities of public network operators. Chances are, however, that deregulation will greatly extend this stated objective – provided the railways bureaucracy has the vision to make Hermes a major force in telecommunications.

With deregulation, a more formidable competitor to telcos will evidently be the private enterprises behind cable television (CATV). Community antenna television networks are examples of broadband structures characterized by an asymmetric line discipline.

CATV brings to the market-place a number of challenges. The most important is that it provides an elegant way to drop the PTT monopoly. With this comes the opportunity to renew POTS. But there are also a number of commercial and technical issues to be solved.

- The proportion of the homes receiving television signals through CATV networks varies from one area to another.
- This happens over a much wider range than that of homes having direct access to the telephone network.

Where cable networks are widespread, advantage can be taken of them in many ways. For instance, by adapting them to support all of the services provided by classical telephony, extending this support to interactive multimedia. Or, transforming them to become part of a capillary structure free of twisted pair.

The CATV business perspectives are enhanced because advanced digital transmission systems on optical fibre cables are being installed extensively. They capitalize on the fact that synchronous digital hierarchy allows flexible interconnection of channels at different bit rates. Many problems, however, still need to be faced and these include:

- End-systems characteristics,
- Dynamic user interfaces, and
- Time schedule for each application.

Other technical challenges include end-to-end protocols, access and network signalling functions, and network termination characteristics. There is also the need to rethink network management functions.

Deregulation has been instrumental in the promotion of new technologies and new solutions. All of the issues we have just seen face all of the operators. Who has the drive to attack them in a forceful manner, and solve them, is the one who wins. The laggards will eventually disappear – no matter if in the past they had 'this' or 'that' monopoly.

THE SYNCHRONOUS OPTICAL NETWORK (SONET) IMPLEMENTATION

The Synchronous Optical Network which is in operation in the United States supports a synchronous frame structure and provides a multiplexing scheme to digital signals. It is a standard for a megastream speed optical network operating at a bit rate of at least 51.84 MBPS – which is labelled OC-1, and is the lowest bit rate in the hierarchy of SONET.

SONET is a non-proprietary protocol. Developed and implemented as an American standard in collaboration with ANSI and CCITT, the SONET project put forward the goal of a unification of the trans-

Figure 4.2 Prevailing dataflow in computer resources

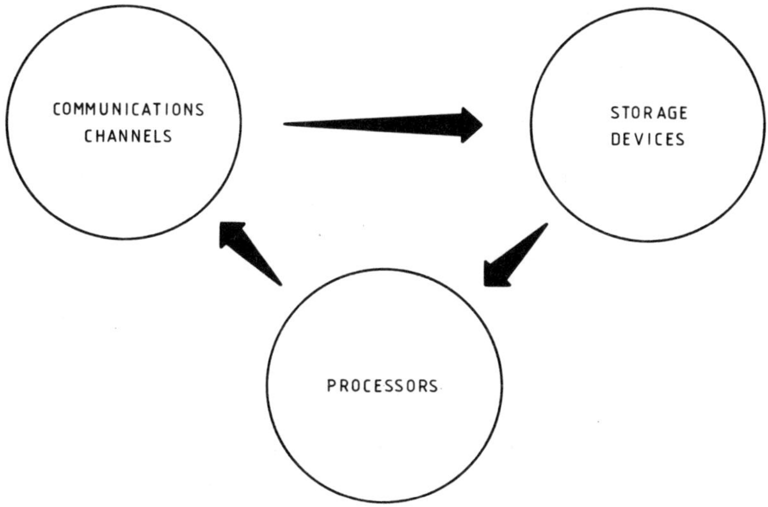

mission system, as well as that of more efficient multiplexing in telephony.

SONET is also seen as a possible solution to help the implementation plans for bringing fibre to private homes. Cable TV and telephone services could be integrated, providing through SONET digital video and broadband connections of up to 44 MBPS on a dial-up basis. Another area of implementation might be the interconnectivity of LANs in competition to FDDI and other standards. We will talk of FDDI in a later section.

In the opinion of many experts, the most important feature of SONET is the possibility of allowing different terminal equipments to work together, bringing all kinds of services to the subscriber through centralized network control. Figure 4.2 provides a bird's-eye view of the prevailing dataflow in computer resources which can be effectively served through the new transmission technology.

A parallel goal is the ability to establish and maintain an open vendor policy. One of the frequently discussed further-out objectives of SONET is to be the groundwork for a multivendor environment within which are effectively used:

Table 4.1 SONET hierarchy of optical carrier levels

SONET Label	*Synchronous Transport Signal Level*	*Bit Rate in Megastreams (MBPS)*	*Rounded-up Bit Rate in Gigastreams (GBPS)*
OC-192	STS-192	9953.28	10
OC-96	STS-96	4976.64	5
OC-48	STS-48	2488.32	2.5
OC-36	STS-36	1866.16	1.9
OC-24	STS-24	1244.16	1.25
OC-18	STS-18	933.12	0.94
OC-12	STS-12	622.08	0.62
OC-9	STS-9	466.56	0.47
OC-3	STS-3	155.52	0.155
OC-1	STS-1	51.84	0.052

- A common bit rate for the optical interfaces, and
- A common frame structure for transmission purposes.

In the line hierarchy of the SONET standard, the abbreviation OC-N denotes the level of the optical carrier. The line rate OC-1, of which we spoke in the preceding paragraphs, is equivalent to 810 channels with a rate of 64 KBPS each (a relic of ISDN; see later sections of this chapter). OC-N is N times 51.84 MBPS. The SONET hierarchy of optical carrier levels is given in Table 4.1.

Personally, I look at the Synchronous Optical Network not as a long-lasting solution, much less one with ambitions of becoming an international standard – but, rather, as a very worthwhile test for ATM packet traffic as well as other packet formats in a WAN.

- SONET allows a dedicated bandwidth to be requested via a virtual circuit.
- The synchronous nature of the network permits the connected systems to predict network performance for the session.

As a synchronous transmission structure, SONET is often used for framing and synchronization at the physical layer of the ISO/OSI model which is shown in Figure 4.3. This is not the level for embedding intelligent network features which can be nicely supplied through what I call the 8th layer over ISO/OSI and in network control.

Figure 4.3 The ISO/OSI model enriched with an eighth layer for seamless network integration and with network control functions

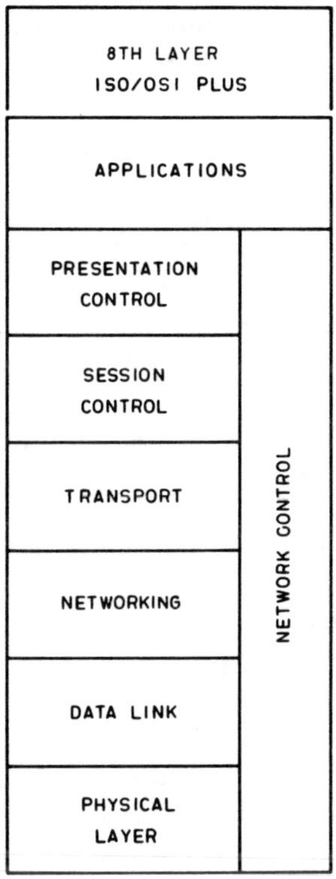

At the physical level of reference the SONET standard defines the optical signal levels, a synchronous frame structure for multiplexed digital traffic, and the operations procedures for the physical layer interface for use in optical networks.

- The basic time unit of a SONET frame is 125 microseconds.
- The SONET format is currently supported by single-mode fibre, multi-mode fibre, and twisted pair.

How is SONET comparing with the synchronous digital hierarchy, which we studied in earlier sections? The answer is more marketing-oriented than technical. After the transmission hierarchy for optical signal levels was established as a global norm, work began on a universal multiplexing and switching protocol to support integrated transport of multirate traffic. But every major vendor looked after his own interests.

In spite of all the talk about international standards, given the fierce competition it is only reasonable to expect that projects along the synchronous line discipline take different paths. Though the common goal has been to support the diverse requirements of multiple-bit-rate traffic while providing flexible transport and switching services at an affordable cost, the solutions which are being advanced today are not compatible – but neither are they contradictory.

NEW DEPARTURES ARE NECESSARY FOR BROADBAND NETWORK SOLUTIONS

No one in his right mind would contest the fact that multimedia solutions need broadband and vice versa. But while the possibilities opened up by broadband communications seem endless, profound changes in demand and supply are putting new emphasis on the need to use advanced communications techniques – not just wider bandwidth.

SONET and SDH have technical merit, though I would think that both have been a good testing ground rather than a major innovation which changes the way in which we communicate. By contrast, there exist other cases which I consider technical horror stories coming out of compromises and ill-defined objectives.

In the background of this statement lies the fact that concepts and goals advanced in the 1980s about broadband and its implementation might have been valid in their time but they got increasingly awkward as technology progressed. Yet, they still prevail among many telcos, even if they rest on hypotheses which are no longer true. The reasons are operational:

- Some years ago, most private networks carried relatively predictable, terminal-to-host data or voice traffic.
- Now, with new applications, communications solutions must address highly bursty environments and unpredictable transmission rates.

Figure 4.4 Path-finding architecture with distributed multimedia databases

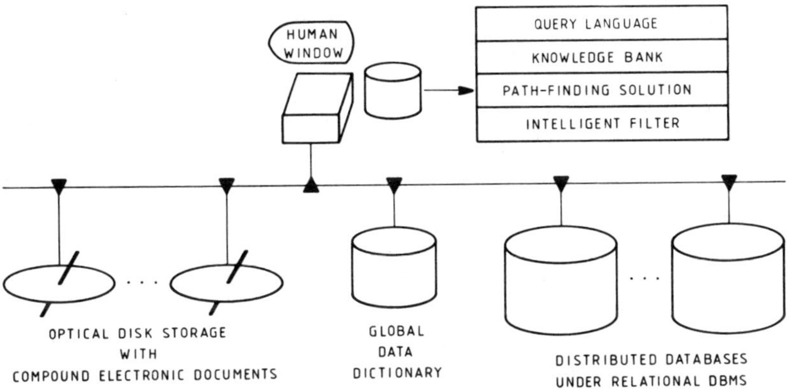

Figure 4.4 gives an example of an implementation environment which is increasingly becoming the frame of reference of modern industry: a path-finding architecture with distributed multimedia databases characterized by bursty traffic. In this context, ideas which might have been valid ten or fifteen years ago have very little if anything to contribute.

Technical solutions able to answer the requirements presented by this environment will be intense in artificial intelligence artefacts – some of them available on public networks but many custom-made to meet specific implementation requirements. However, only a few telcos and the tier-1 financial and industrial organizations have the skills to face these problems.

While some of the solutions being adopted by the more progressive organizations are still hardware oriented, the majority of new approaches increasingly depend on sophisticated software – and most particularly on agents and other knowledge-enriched artefacts, as we have seen at the end of Chapter 3. This is the way to bet for the years to come.

- There are many efforts underway to create a standard for public network management.
- The most advanced approaches focus on agents, or knowledge robots, to serve lines, nodes and clients.

There is little doubt that telephone companies have to face huge investment problems connected to the modernization of old plant, on which they still depend for their services. Upgrading the access portion of the telecommunications network between the customer and the central office is a very expensive undertaking.

- Real technological change cannot reasonably start until customers are served through a modern plant.
- Hence the failure of the half-baked ISDN solution the telcos advanced in the mid- to late 1980s (see also the following sections).

But there is even less doubt that the worst possible solution is that of being paralyzed by the magnitude of the changeover problem. Even if the telcos do not like to admit it, the truth is that it is easier to implement new technology than change attitudes. This is really every organization's weakness.

New attitudes are necessary with the introduction of ATM and SDH in complex networks and their management systems. Network management is so complex because of the prevailing multi-vendor environment.

- It is reasonable to expect that the market will continue being fragmented, even after the standardization effort takes root.
- At least at the application level, the way to bet is that the user interface will still be customized by the vendor.

This I see as persisting in spite of the fact that there is a growing appreciation that one of the crucial factors for return on high-technology investments is knowledgeable and efficient use of standard interfaces. Able solutions will help to increase reliability, respond more rapidly to competition and reduce costs in the telecommunications networks. But who will provide the able solutions?

Absence of an answer to this query leads to technical horror stories. They develop because classical approaches to network design see to it that either circuits are underutilized or users do not get the service they expect. Yet, as customers become more sophisticated and more demanding, the technology and tools needed to deliver communications services must keep pace with new requirements.

THE WISDOM OF KEEPING COUNT OF DOLLARS AND POUNDS

Let us return to the fundamentals. Through the 1980s switching equipment vendors seized upon a single slogan in their bid to capture the market for digital switches: *Integrated Services Digital Network* (ISDN). The conventional wisdom was that by capitalizing on the existing telephone infrastructure ISDN could meet most present and future telecommunications needs.

- This approach, however, has rapidly been overtaken by events: from technology leaps to user requirements, and
- Today's telephone network is in a state of flux with at least three countries: the US, the UK and Germany developing their own, incompatible, wide area solutions through broadband.

Because of technical shortcomings and elusive business visions, the enthusiasm about ISDN has been short-lived. The scheme has been overtaken by technology, and marketing response was so lacklustre that most vendors still lick their wounds.

ISDN works at 144 KBPS (2 × 64 KBPS + 1 × 16 KBPS). That is narrowband, and with narrowband the main market is substitution. Hence a limited and not so rewarding perspective. Another key impediment to the business chances of ISDN has been that the telcos kept their high tariffs.

But while, if the tariff is right and the state monopoly permits it, ISDN may serve a market niche, it will not answer other business needs. The German publisher Gruner & Jahr uses the Bundespost's forerunner Wideband Network (VBN) to transmit text and graphics at 140 MBPS. This is three orders of magnitude more than is supported by ISDN. Who cares about kilostreams under current technology conditions?

If we persist in using narrowband, just because of obstinacy, then the data volume must be reduced – and with it competitiveness. By *how much* depends on the task at hand. For example, if we want to transmit moving video via the 64 KBPS telephone lines of the narrowband ISDN, then we have to reduce image data by a factor of 2000.

- From 140 MBPS,
- Down to 64 KBPS.

Technically we can do that, but we cannot achieve it without losing some of the original information, and a certain amount of degradation is inevitable. Do we wish to pay that price? For what purpose?

Let us never underestimate the impact of dollars and pounds involved in any product offered to the market. Costs eventually hit the bottomline and, therefore, they influence management's decisions.

There is another major dollars and pounds reason for not doing silly things. Companies must maximize their income. The cost of doing business in the public switching industry gets higher.

- New digital switches are roughly 100 times as expensive to develop as electromechanical ones.
- Because of deregulation the chances of success in the market through the use of old technology are getting slimmer.

Suppliers that once counted on ISDN as the 'final step' in meeting telecommunications needs, never figured on the impact of *intelligent networks*. Yet today intelligent networks are fast becoming a whole new battleground, and winning meaning having ready:

- The processes,
- The policies,
- The tools.

In the evolving environment of networked services, common carriers are faced with the problem of how to upgrade their switching base. The issue is a troublesome one, as each telecom has thousands of switches already installed and major upgrades must be made before new services are offered.

For instance, in its Network Systems 5ESS switches, AT&T handles ISDN upgrades in groups of 512 lines. But non-ISDN 5ESS switches must be fitted with one or several Integrated Services Line Units (ISLU) – or, rather, non-integrated. (The 'I' in ISLU is a misnomer under the technology and user requirements of the 1990s. The same is true of the 'I' in ISDN.)

Poor AT&T. Beyond the ISLU it is necessary to install a packet-switching unit with ISDN software into the 5ESS main module to handle ISDN 'D' channel signalling transmission. Money spent for nothing because of misjudgment. Companies are made of people and people never learn. Costs should always be examined under the aspects of:

- Utility,
- Performance, and
- Functionality

which is provided by an investment. Under this reference we can do better than ISDN by an order of magnitude in terms of cost-effectiveness, and this with state-of-the-art technology.

With a host of technological platforms to choose from, one of the great challenges facing telecommunications manufacturers today is anticipating the needs of global markets while preserving their ability to offer individualized solutions for different national networks.

- The requirements of global markets evolve very fast, and
- The need for channel capacity grows almost exponentially.

From banking to manufacturing and marketing, fast-developing user demands in the communications domain create a first-class business perspective but also market instability. This is compounded by the relative decline in the importance of hardware switching, as its functions are integrated with other network elements such as transmission systems and computers.

Solutions should look to the future rather than the past (the twisted pair) in order not to paint ourselves into a corner. The face of the public network is transformed by:

- The advent of the intelligent networks, and
- The increasing competitiveness of fibre optics.

Beyond that, the intelligent network blurs the boundary between public and private networks through the provision of *virtual communications facilities*. Correspondingly, the deployment of fibre optics and intelligent electronics in the local loop alters the boundary between switching and transmission.

With the use of intelligent multiplexors and optical fibres in the local loop, flexible access systems should deliver true customer control at *low cost*, and with almost *unlimited bandwidth* on demand. Coupled with the continuing growth in private networks and the explosion in cellular and cordless mobile communications, it is now recognized that ISDN will only be one of a number of telecom infrastructures – and not the most profitable one for that matter.

PUTTING 52 MBPS ON THE COPPER LOOP. FOR WHAT PURPOSE?

Strange things happen these days in the silly effort to revive old equipment whose life has already been spent. After the lame duck ISDN we now have a new brilliant concept: the 52 MBPS to be supported on the obsolete copper local loop – at a great cost and for no real purpose. That is at least what some telephone people are now saying:

- The twisted copper pair is set to get a new lease of life as a broadband pipeline, and
- Some vendors hope to push its capacity to ATM speeds, no matter how bad the bit-error rate may be.

That this sort of talk comes from two start-ups, Amati Communications and Orckit Communications, is not surprising. What is suprising is that AT&T also harbours similar thoughts about developing approaches that could give carriers up to 52 MIPS over existing copper cabling.

Orckit, for example, is working on a broadband asymmetric digital subscriber line. The set-up consists of a high-speed modem at the customer's premises and network-based loop carrier software. This scheme is claimed to be offering:

- 52 MBPS downstream, network-to-user, and
- 6 MBPS upstream, user-to-network.

Let us not lose sight that here we are talking of old plants – some of which were installed at the beginning of this century, and should have been decommissioned long ago to make way for advanced communications solutions.

The projected cost for this abortive concept is hair-raising. Initial shipping prices are estimated at $2500 per subscriber (*Communications Week International*, 1995). Though that price is expected eventually to drop, nobody seems to be asking the critical question: 'Who would be silly enough to pay $2500 as an entry price to old plant?'

- Reliability and BER seem to have taken the back seat, and business opportunity is on cloud 9.

- All this is proposed in the name of multimedia – as the current prototype's speed ties to ATM.

By some miracle, unreliable twisted pair is said to be answering the telecom operators' plans for video on demand. Among its many other shortcomings, this proposed questionable technology has a distance limitation of 240 metres, though by cutting the speed to 26 MBPS the reach seems to increase to 600 metres.

Many technical problems go unanswered. One of the bigger difficulties is getting into the home: whether the reach is 240 metres or 600 metres, the last 100 metres will be most complex and expensive – this is at least what the 'new technology's' fans are saying.

- The hypothesis is that with their huge embedded copper bases, most telephone operators are looking at any method that unlocks the bandwidth of old plant – no matter if it is sound or silly.
- But the worst way to go after solutions is to have good money run after bad money which happens to be embedded in old, low-quality plant.

The developers' best hope with this new technico-economic abortion is that in many telcos and other PTTs the bureaucracy is so hooked to the past that it may end by adopting the beef-up, twisted-pair scheme. It is however to be seen if the customers who count will buy the service and pay for it.

The story of this 52 MBPS technical miracle is not much different from that of videotext in the late 1970s and ISDN in the mid-1980s. In my book '*System Architecture and System Design*' (1989), I commented as follows on the six deadly sins of ISDN and its bleak longer-term business perspectives.

1. Industry requirements projected for the mid-1990s stand at 2 to 3 GBPS.

This is more than 4 orders of magnitude greater than the combined 144 kilostream of ISDN. There is also a low BER undermining the ISDN initiative because:

2. A very important issue is transmission quality and dependability.

Expressed in BER, voice-quality lines range from 10^{-2} to 10^{-4} and 10^{-5}. The so-called 'data quality' lines have a BER of 10^{-7} and 10^{-8}. All this is old plant. What is *really* needed is 10^{-11}, 10^{-12} or better. Old plant cannot provide it.

3. Integrated services packet networks call for a top-down grand design and a bottom-up detailed design. Not patches.

By addressing itself only to the local loop and its obsolete equipment, ISDN is a patching approach. It is not a new solution.

4. In the hurry to get the patchwork going, telcos and PTT forget they would have to change switches and trunks.

Eventually, this will mean double investment rather than savings on current aging plant. Economically, ISDN does not make sense.

5. Even the too little, too late of the ISDN recommendation is applicable only to a very limited topology: the local loop's connection to the city office.

Still, nobody can guarantee the recommended 'standard' will be applied in uniformity. We have had that experience with X.25.

6. ISDN offers a 2B+D wire (2 x 54 KBPS+16 KBPS return wire) but does not provide for multiple session capability – which is crucial.

Splitting the session is fundamental to enable workstation utilization. It must be done either through an X.25 node or a front-end. Current implementation experience indicates that neither is performant.

'Even human stupidity', one of my professors at UCLA used to say, 'has its limits. And where stupidity ends the conflict of interest is beginning.' Sure the telcos have a conflict of interest. The point, however, is that – like ISDN – 52 MBPS on twisted pair is not an event which is part of the solution. But it is part of the problem.

A GROWING LIST OF ISDN NEGATIVES

Since 1990, Intelsat and Eutelsat have been pressing claims that standards developed for ISDN are in many cases not compatible with satellite services and could seriously impede their use. These claims by the worldwide satellite organization and its regional European counterpart have exposed a fundamental design flaw in the integrated services digital network.

Working on unfounded hopes and in a superficial manner rather than doing their homework, the PTTs, other telcos and telecommunications manufacturers have spent a lot of money in developing, testing and producing something for which there is no apparent profitable market.

- In retrospect, ISDN indeed turned out to be the last of the telecommunications white elephants.
- This contrasts very badly with the original claims that it was going to become the 'next stage' in the evolution of global telecommunications.

Many lessons can be learned from this failure. Conceived in an era of monopoly in telecommunications, with the primary goal being a better use of old plant, the ISDN concept became an anachronism in the era of network pluralism, competition and new technology.

To its proponents – and there have been many among the world's manufacturers, public network operators and light-headed consultants – since its conception ISDN represented a great breakthrough. To its critics, ISDN was nothing but a miserable way to access and use the global telephone network.

Based on the limited capacity and high bit-error rate of the wired telephone established in the late 19th century, ISDN promised worldwide connectivity for voice, data and image communications. It also advertised a common infrastructure for open communications extending the Open Systems Interconnection concept to multimedia – and other nonsense. Only open-eyed industry watchers appreciated:

- That apart from its other deficiencies, ISDN is a pure voice digital network.
- It is not multimedia – yet multimedia solutions are the future.

- And it is far from being an integrated network in a cross-country sense.

The weaknesses did not take that long to detect. A decade or so after ISDN development work first stepped into the spotlight, there remains a deficiency of applications and the major user organizations in the telecommunications arena have been turned off. There is absolutely no mass market – as for instance there is for personal computers.

- For one thing the PC industry never had the approvals, availability and standards problems perennially faced by ISDN applications developers,
- And for another, ISDN is the (stillborn) baby of bureaucrats, not of the free enterprise.

As the years passed by, the lack of applications left most corporate users skeptical of whether ISDN can enhance their communications capabilities. No wonder there has been very little genuine demand for it.

Advances in intelligent network and broadband services in the public realm, made people and companies wonder if ISDN will ever be the star its proponents spoke of. A high-speed fibre-optic link to the south of England, for example, allowed facsimile and electronic mail to be transmitted 25 times faster than over an ordinary phone line, and videoconferences were charged at the same rates as ordinary phone calls.

People with the salt of the earth in their judgment cite incompatible ISDN terminal adapters and PC cards as another problem for applications developers and end-users. If ISDN was not hampered by such standards and rollout problems, companies might have placed more emphasis on developing applications.

- But there are currently so few applications, and appropriate equipment on customers' premises is so sparse, that users have little incentive to connect to ISDN.
- Big-headed equipment manufacturers are ignoring the needs of small and medium-sized businesses by not developing small ISDN PBXs and key systems.

Neither have major manufacturers put their full marketing clout behind the ISDN products they have released – as if they don't really believe in them. Products are emerging and stay stagnant in terms of market penetration.

At the same time, different implementations of ISDN standards have hindered the availability of the product. For example, because public switches apply ISDN differently, companies are unable to market their product internationally.

As the years passed by and money was spend without any return-on-investment, the truth dawned on the ISDN enthusiasts that the possibility of using the old plant for new and demanding applications – as well as of using the same terminals and standardized services across international frontiers – was a chimera. In each of these areas ISDN has been a comprehensive failure.

As the sobering annual progress report on ISDN started pointing out, users are still doubtful of its possible achievements, and the projected market has been a mirage. Corporate customers have already turned to new or more readily available local, metropolitan or wide area network technologies.

- The results have hit head-on precisely those manufacturers and operators who were responsible for hyping the ISDN slogan in the first instance.
- Divergent policies and incompatible plants led to implementing national versions of ISDN that are now heterogeneous on an international scale.

In conclusion, the reality is that ISDN cannot even evolve as a core infrastructure network; it will be only one of a number of digital access technologies to the public network. The enthusiasts of the 52 MBPS twisted-pair balooney should taken notice. Let us hope that the hype does not enter the Information Superhighway and the same mistakes are repeated all over again – but on a grand scale.

THE FOGGY BOTTOM OF B-ISDN

Not all projects advanced by telcos and PTT organizations bear fruits. This is particularly true of those based on old concepts. The example is not only ISDN but also the so-called *broadband ISDN* (B-ISDN) which may become an anachronism before a single switch has been installed.

Everywhere competition is spreading, and as commercial realities take hold some hard questions are being asked about the vision of a ubiquitous, homogeneous broadband network which nobody really believes is doable. Telecommunications operators who have been expected to construct a global broadband integrated services digital network are:

- Being frustrated by the near-stagnant demand for their main product, the plain old telephone service, and
- Face a current or near-term prospect of sharply falling tariffs and prices – which hits their bottomline.

Within this environment, B-ISDN has been stillborn since its conception. It was thought to be a single network capable of handling every imaginable type of user traffic, including voice, data, text and image. Its central element was supposed to be the asynchronous transfer mode switch, but only lip service has been paid to universal standards.

Following the euphoria which followed the advent of another dead duck – ISDN – a preliminary agreement was reached on broadband ISDN in 1988. Work on more detailed standards is always placed sometime in the future. Because of the involvement of the International Telegraph and Telephone Consultative Committee (CCITT), few people at the time took notice of the fact that the conception of broadband ISDN does not fit neatly into today's telecoms environment, with its:

- Highly competitive markets,
- Unpredictable traffic, and
- Multi-faceted user networks.

Many telecom experts suggest that a more plausible model is a world in which a few common broadband transport networks are shared by many different service providers and systems integrators. This can be done by using a *range* of routing and switching technologies with variable but high throughput.

Other communications experts argue that the problem of falling prices for existing services has destabilized all telecom operators and they no longer have the patience to plan for an advanced broadband network which observes a universal standard – a strategy beset with risks.

- ISDN was conceived as a replacement for the public switched telephone network, but it has been half-baked resulting in an array of fairly incompatible solutions.
- For most carriers, the spectre of the failure with narrowband ISDN looms as a grim warning of what can go wrong with the B-ISDN grand schemes.

In the case of broadband ISDN the risks for telecoms appear to be even greater than those borne in ISDN. Broadband services priced at competitive levels would slash revenue from bread-and-butter voice services that currently provide at least two-thirds of their revenues.

Broadband at reasonable tariffs implies a telephone service much cheaper than that available today, particularly in deregulated environments. By contrast, in a regulated setting high-quality broadband capacity would not be affordable with users condemned to very low quality like that of the plain old telephone service – as we saw in an earlier section.

In fact, tier-1 telephone companies are commenting that in a deregulated broadband landscape POTS services would be practically free. Hence, the telecom operators' core revenue will no longer come from voice-grade telephony. This poses the second problem in the realm of B-ISDN:

- Are the different telcos ready to provide sophisticated, value-added services?
- Do they have the concept? the products? the technical skills? the ability to compete?

Let us not forget that with the exception of the US and England, the other ex-PTT outfits still remain huge bureaucracies which – like banana republics – have more good-for-nothing generals than skilled people. If it was not for the high tariffs the government permits them to apply, they would have gone bankrupt – which one day they may do with or without broadband services.

Finally, the one thing that might have made sense with B-ISDN: a universal, cross-country standard does not seem to be doable. This is not so much a matter of will, but of technology. No telco or other PTT can afford to reinvent everything, but the more they rely on bits, pieces and big chunks of old plant – like they did with ISDN – the more they condemn themselves to the dust of history.

WHAT MAY BE THE PROSPECTS OF FDDI?

Fibre Distributed Data Interface (FDDI) is a LAN standard for a fibre-optic network with megastream data rate. FDDI-I featured 100 MBPS; FDDI-II, 150 MBPS. Both specify a dual-counter rotating-ring topology, but neither is a typical implementation of fast packet-switching.

With the capacity of 100 MBPS, FDDI-I is a token-ring protocol designed mainly for LAN and MAN applications. The problem is that an FDDI packet does not map efficiently into many of the WAN architectures. Furthermore, the bandwidth gains have not made FDDI a viable option for most users.

FDDI-II, also known as FDDI Follow-on (FFOL), has been proposed as a solution to some of the problems with FDDI – with special attention given towards realtime multimedia capabilities. The FDDI protocol itself took an inordinate amount of time to complete. But:

- The FFOL protocol is still in the design stage, and its arrival may come too late to compete with ATM.
- This reference is quite similar to that applying with B-ISDN, another hyped 'standard'.

In theory, through the FDDI protocol, with a multimode fibre-optic cable it may be possible to have a total cable length of several kilometers per ring. This is important in using this approach as a LAN backbone with users connected to the FDDI through Ethernet or other LAN. But in practice there have been many shortcomings.

Capacity is another key issue. In the mid- to late 1980s, when FDDI started being promoted by vendors, the 100 MBPS were seen as a respectable transmission speed – ten times faster than Ethernet. But:

- Today 100 MBPS is far from being an exciting solution, even if it were technically perfect, since

- Troubles plagued the first offers and as a result both vendors and users became cautious.

A critical examination of FDDI has come from users and suppliers trying to maximize existing technology without running into weaknesses, such as vulnerable cabling or complicated management. A very careful approach is recommended because, in an implementation sense, FDDI has generated more hype than any other standard for computer networks.

- The offer of 100 million bits of informatiopn per second transferred from one computer to another across 2 kilometers or more of optical fibre has been the hype.

At such speeds, a Unix workstation could dump its entire 200-megabyte hard disk onto another workstation in a mere 16 seconds.

- But FDDI bugs left aside, computers and communications software cannot always handle such volumes of data so easily, without high performance protocols – and, in cases, customized solutions.

There are also critical flaws in FDDI implementation that limit the number of attached devices on a ring. For instance, for systems using an optical switch to bypass inactive nodes, the signal becomes too weak for transmission if more than three consecutive devices are out of action.

Another problem has to do with bridges and concentrators. Two techniques for linking Ethernet with FDDI, for example, are encapsulation and translation. With encapsulation, the bridge takes Ethernet packets and wraps FDDI addressing information around them for passing on to another bridge. Because there is no standard for encapsulation, any implementation is essentially proprietary.

Granted, encapsulation has been superseded by translation, where packets are converted from one format to another using a standard technique. But translation has other problems such as delays – and the interfaces are not normalized. Bridges, encapsulation and translation are necessary because FDDI is not suited for a direct connection of workstations:

- A workstation which has failed might interrupt one of the rings of the FDDI dual ring system.

- To prevent this from happening, an optical bypass switch must be installed for a minimum protection of the ring.

The dual ring topology is capable of facing the single interrupt, but FDDI is relatively sensitive to interrupts even if a single interrupt does not break down the whole network. Hence the whole fibre-optic cabling must be evaluated very carefully. Also the distances between adjacent nodes must be known exactly to properly choose the right type of fibre.

Another reason why FDDI did not take off when it still had a chance of doing so, is that right after it was announced came the shadow of its own replacement – FDDI-II. The premise that with this standard it would be possible to transmit isochronous traffic like voice and video signals – if and when it gets market acceptance – has still not been fulfilled.

There is also the so-called Fibre Distributed Video/Voice and Data Interface (FDVD). With this 'standard' it should be possible to integrate isochronous services and to get a better compatibility to B-ISDN – provided that one day it comes to life. The list of hypes continues; and so on and so forth.

If one says there are too many standards thrown into the kitchen sink and the whole situation is becoming ridiculous – he or she would be right. But for the standards fans let's add that there is also the Survivable Adaptable Fibre-Optic Embedded Network (Safenet). Networks based on these different and incompatible standards provide confusion rather than services – in spite of the fact that there is a crying need for solutions able to serve new-generation multimedia realtime systems.

Part Two

The Race for Multimedia Servers and their Implementation

5 Business Perspectives of Interactive Multimedia

INTRODUCTION

The invention of new technologies, as well as new materials, can signal the birth or the extinction of an entire industry. For instance, silicon chips decimated the vacuum tubes business, and optical fibres are rapidly replacing the telecommunications market for copper wire.

- Because of supporting the display of many types of information – like moving images and animation, along with the more classical voice, text and data – *multimedia applications* displace older technologies.
- This substitution is taking place not with a big bang, but at a relatively rapid pace. Hence, the rush to mergers and alliances of which we spoke in Chapter 1.

Many companies invest in advanced technological solutions in the hope of developing new and profitable businesses, but laboratory work alone does not suffice. A company must succeed in *commercializing* the invention through rapid time-to-market. Multimedia, the core issue of Part Two, is no alien to this principle. One of the most important obstacles in a rapid time-to-market approach has been the failure to recognize:

- How long commercialization is likely to take, and
- How much it is going to cost not only in money but also in human resources.

In the past, there has been a long interval between invention and widespread adoption of a new material; or the development of a new process and its popularization. But during the last quarter century this delay is getting shorter and shorter, making it very difficult for a company to recoup its investment if it fails in the race to the market.

The synergy now sought through the merger of media companies and joint ventures with software firms, promises to further accelerate this process. At the same time, the ongoing market globalization requires a close collaboration on an international scale, and this poses problems for cross-border operations.

Cross-border multimedia applications are able to position themselves more competitively, and get better acceptance, if globalization is seen as an integral part of *productization* – and vice versa.

At the same time, new developments can significantly improve their profit potential, if protocols and bandwidths are standardized. Heterogeneous line disciplines and transmission codes are a hindrance in multimedia ventures.

Finally, the quality of the management team is crucial. Even the best products cannot win if management does not provide research, development, production, marketing and distribution *leadership* – and promote an agile, comprehensive support for innovative and appealing customer services.

MARKETING THE NEW MULTIMEDIA BUSINESS VENTURES

The Introduction underlined that both the development of novel networks and imaginative products, and the acquisition of management skills, are crucial to success with multimedia. The same is true of *standardization* – and there exists historical precedence to document this reference. A lesson can be learned from *penny mail*; more precisely the effect of uniform letters and handling costs in 19th century England:

- Prior to uniformity through normalization, 76 million letters were written in 1839
- But after normalization of letter sizes, 169 million letters were written in 1840 – more than doubling the market size.

Standardization has not yet happened with the multimedia business and this industry is getting into shape in a way quite different that what was thought would happen even three years ago. The drive is towards concentration in a rather small number of giant firms. The August 1995 takeover of Capital Cities/ABC by Disney pressed the point that today entertainment is:

- The principal American export, bigger than aircraft,
- A new-job generator, more important than health care, and
- A technology driver; a developer of more sophisticated software than Star Wars.

Through a process which can be seen as the direct opposite to standardization, entertainment capitalizes on the fact that after the end of the Cold War the motor behind the economy is the consumer's object of desire which accounts for:

- More than national defense,
- Even more than housing or cars.

Disney provides an example. The first instance was the announcement of 31 July 1995 that it was buying Capital Cities/ABC, the owner of a leading television network, for $19 billion. Then, on 14 August, Michael Eisner announced that Michael Ovitz would join Disney as its new president.

A dealmaker rather than a rocket scientist, Ovitz, who has been his own boss for 20 years and is often believed to be the most powerful man in Hollywood, will be responsible for Disney's operating division. This includes filmed entertainment, consumer products, theme parks and Capital Cities/ABC – the bottomline being diversification rather than standardization.

The Ovitz reference is important because it tends to suggest that a different pyramid of *strategic planning* values will prevail in the multimedia business than that which has characterized the post-World War II years. Figure 5.1 explains this reference and brings the new values into perspective:

- At the base is market understanding and the evaluating of strengths and weaknesses to be covered through acquisitions.
- This is followed by strategic planning, and since the consumer market is cost-sensitive by cost-cutting efforts.
- Programme launching requires the so-developed infrastructure, but market penetration calls for significant training.

Figure 5.1 Strategic planning is a polyvalent enterprise – from market perception to sustained profits

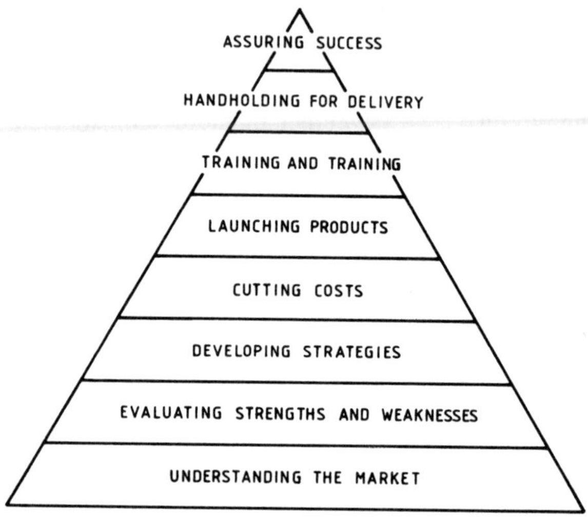

Without any doubt, hand-holding is a prerequisite to the assurance of business success – a fact which is valid with all new technologies, not only with multimedia. But multimedia has another major asset that other business does not necessarily possess: the financial facts characterizing this industry are startling due to the fact that:

- Publishing, cable distribution, TV programming, film and the telecommunications business are becoming one distribution system.
- The market for entertainment and information is growing much faster than the multimedia ventures are consolidating.
- The deconstruction of old market segments by technology is quickly being followed by new agglomerations.

The race is on for media companies to control distribution channels in broadcast, cable, film and telephony. Technology which can be effectively marketed to the consumer is the driving force chopping and mixing once-discreet industries into new configurations of consumer services with a global impact.

Some securities analysts now believe that this may be the first true manifestation of the Information Revolution sending a number of corporations towards one another in search of synergy, while at the same time expanding the total economic universe within which all media companies play.

As we have seen in Chapter 1, telecommunications companies are as much in the multimedia arena as are film studios. Some Wall Street analysts think that the September demerger of the old NCR/Teradata computer business and the ex-Western Electric from AT&T was just the preliminary step for the rejuvenated Ma Bell to enter the multimedia business.

AT&T had no choice but to make a bold move. MCI, America's second largest long-distance telephone company, has been investing $2 billion in Rupert Murdoch's News Corporation, which owns Fox, the fourth American TV network and a major Hollywood film producer. The theme is to use News Corporation's content for networked consumer services.

- This sort of deal is the reality of convergence in the multimedia markets.
- Odd couples from media, computers and telecoms are getting together for synergy and greater market appeal.

Both established companies and media investors are on the move – and they sense lots of potential profits. Barry Diller, the man who built the Fox television network for Rupert Murdoch, said that he had reached a deal that would give him control of the sixth-largest US television station company, Silver King Communications. Silver King reaches 28 million American homes, and could form the backbone of a new nationwide television network with cross-border ambitions.

Prior to following up on the business perspectives of multimedia, which is the main theme of this chapter, it is wise to take a quick look at some technical aspects. This is the theme of the following section.

TECHNICAL AND TECHNICO-ECONOMIC ASPECTS OF MULTIMEDIA

The first important technical characteristic of the multimedia era is *bandwidth*. We have spoken of bandwidth in Part One. It is the

Figure 5.2 Projected bandwidth utilization with multimedia applications

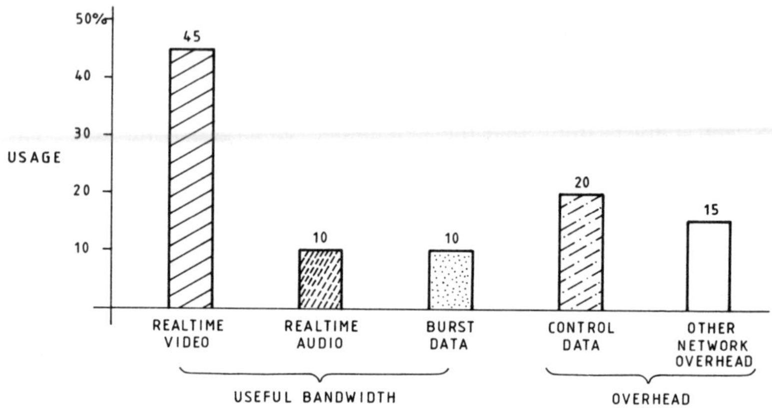

range of frequencies characterizing transmission performance and its limits. It is measured as bit rate in a signal line, which represents the speed at which information flows.

What has not been explicitly said so far is that, as a term, bandwidth also has semantic and information content. It is a metric of the capacity of a network connected to the other crucial reference: the *bit-error rate* (BER), which characterizes the quality of the line. Part One also underlined that, in terms of bandwidth, we distinguish kilostreams, megastreams and gigastreams.

- Figure 5.2 presents projected bandwidth utilization with multimedia applications, distinguishing between the useful band and overhead.
- As it can be appreciated, realtime video will probably take 45 per cent, the lion's share, while the two overhead functions will consume an impressive 35 per cent of capacity.

Multimedia applications need megastreams and gigastreams, and this large information flow capacity has to be managed. Realtime management, optimization and diagnostics will be done through software. Next to bandwidth, *software* is the most important technical factor which can make or break multimedia opportunities – which helps to explain Microsoft's involvement in multimedia ventures.

The third most important priority is the hardware gear and software–hardware modules required for agile system solution. Each major applications domain will have its own specific requirements – but some components will be common, at least in terms of functionality if not in regard to the particular units. Taking the home workstation (HWS) as an example, the systems solution will consist of a number of extensions:

- Modem with polyvalent capabilities,
- PC and slot,
- HiFi home network,
- Digital receiver (Set-Top Box) (which, as we will see, could be substituted by the PC and software),
- Video recording, and
- Video-on-demand.

Superficially, it seems that many of these modules are already in the home. In the general case, however, this is heterogeneous equipment which does not work well together as an integrated system. Hence, while the need for standards which has been underlined in the previous section may not be appreciated by media companies, it will be a crucial element when it comes to user systems.

The best solution I can see is a software-supported modem polyvalence. This is supported by the fact that the multimedia market is characterized by diverse services – from the Internet and Compuserve to the new ventures we have seen in Chapter 1 and in this chapter. These will compete for a variety of offerings, with homeshopping, homebanking and video-on-demand being only three examples.

Once they enter the home by means of the broadband channel, service providers will try to build their business inside the consumer's residence. Examples are the online interconnection of fax and videoconferencing services, all the way to stereo home networks and paying 2-way TV channels via smart cards.

The keyword in all this is *innovation*, which constitutes one of the most important priorities, bringing research and development to the frontline. But there exist drawbacks because competitiveness in a mass market means not just innovation but also *cost control* – and this has not yet become a wide conscience. Neither have the multimedia companies yet demonstrated their ability to merge and align:

- Technical activities, and
- Business strategy.

A different way of making this statement is that in the developing multimedia business, leadership must be both technical and economic. Company management should comprehend that at an annual interest rate of 8 per cent, a dollar invested today must yield a five-fold return 20 years from now for investors to break even.

There exists a host of other issues relating to technical activities, some of which we will examine in later sections. But for the individual company competing in this market, of the two points presented above, my choice is to give priority to business strategy. After all, in a well-run organization business strategy should always precede the answers to be given to the technical problems.

ALIGNING TECHNICAL ACTIVITIES WITH BUSINESS STRATEGY

One of the basic reasons why it is so important to align technical activities with business strategy is the novelty of the multimedia market. In the last couple of years, there has been an explosion of new companies whose business focuses on the ability to merge telephone and television services, with an eye on positioning to compete in the information superhighway. American firms have taken the lead and some have decided to use England as a test market.

This is the case of TeleWest Communications, a provider of residential and business cable-television services. Operating in the UK, it is a joint venture between TCI Communications, the American cable-television operator, and US West, one of the Baby Bells, the regional telephone companies in the United States.

Another company, Nynex CableComms, provides cable TV and telecommunications services to an area of nearly 3 million British homes. It is two-thirds owned by Nynex, the major American regional phone company which operates in New England. The remaining shares are listed on the London Stock Exchange and on NASDAQ; but investment analysts have some queries:

- Are British consumers really interested in so-called interactive services that marry multimedia technologies?

- Can telecom companies or broadcasters provide these services with the synergy of partnerships?

In other terms, how sound can these investments be? The answer is that if things go according to plan and they get anything like a market penetration of the 45 per cent they are projecting, these companies will be wonderful investments. If they stay stuck at around 21 per cent, as the case is now, they will be poor performers – and the latter is more likely than the former.

However, the signals which could provide a factual and documented answer to this query are still scrambled. As with Prestel, the 1979 videotex service of British Telecom, market penetration so far is not good. Statistics are stuck at around 20 to 22 per cent of British households that seem willing to subscribe to such services. This is too low a level to foretell eventual profitability.

There is no lack of reasons for the consumers' cool response. Some financial analysts think that programming is not really top class; and that if it were improved consumers could be convinced that cable represents good value. Then, it is being suggested, market penetration might increase to between 35 per cent and 45 per cent.

My thinking is different. Learning a lesson from the failure of Prestel, and other similar ventures, multimedia is not for the consumer – at least not for now. It is for *the industrial corporation.* This choice is a crucial element in making the multimedia drive a business success or keeping it a failure.

While multimedia opportunities in the consumers' market do exist, I am personally convinced that the greater potential for profits is on the trading floor and the board room rather than with home electronics. Multimedia solutions will be valuable to decision support systems in the process of:

- Protecting one's flanks in the business market,
- Sustaining cost-effective teleconferencing, and
- Managing risky products like derivatives (see the later section on financial applications).

The choice whether the business of a multimedia company is best served through a mass-market or a corporate orientation is *strategic* – and it is not easy to do. Once done, it conditions the whole enterprise because, as Figure 5.3 suggests, it impacts on an array of:

Figure 5.3 Selecting the market target is indivisible from providing advanced technical features

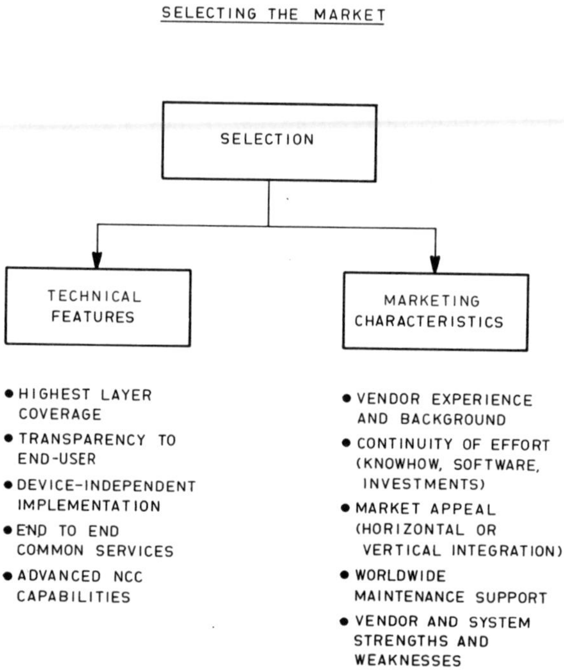

- Technical features, and
- Marketing characteristics.

This choice also has aftermaths on many policy issues. As already discussed, the crucial competitive element in a mass market is *low cost*. By contrast, in the corporate multimedia market *innovation* will play the most critical role.

The Electronic Journal project of MIT which was addressed at the end of the previous chapter, further elaborates on this market segmentation. It does so by examining the project's findings that there exist not two but four market strata. But as the reader will see in due time, these add up to the same bifurcation: the business versus the consumer market.

A business market orientation leads back to what we said in the last section about research and development, and the benefits we can derive from such investments. As with other branches of industry, a vital factor affecting return on investment from R&D is the ability to provide management with imaginative solutions. These new solutions must be cost-effective and able to be generalized, in order to sustain return on investment – keeping in mind that:

- All new processes are likely to be inefficient in the early stages of the learning curve and their yields are relatively low.
- This forces developers to keep prices high to recoup their investment, which ends by being counterproductive at the marketing end.

In the general case, the longer it takes to bring down the costs of a new technology the slower the commercial payoff will be, and the less time will be left before the patent on a given invention expires. The solution is for multimedia companies to invest more in the process of generalizing a new solution as in developing it – and the same is true of the new product's marketing budget.

Investments in research, development and marketing should be tuned to speed commercialization and implementation, increasing the innovation quotient, improving reliability and assuring customer hand-holding. Yet, studies show that industry in general gives far too little attention to implementation, not only of multimedia but as a general rule. In America and Europe,

- 70 per cent of R&D funding goes to product development (including basic research), and
- 30 per cent to process development, with an insignificant amount going to marketing.

In Japan, these percentages are totally different – and the same is true of the software industry at large. Since process technology and marketing will most likely be the key to a company's success in commercializing multimedia products, top management has to re-evaluate the time and money spend on development. In particular, it should foster a cooperative research effort in the early stages of the process of innovation, freeing up the resources needed to focus on customer hand-holding and marketing.

TECHNOLOGY SCOUTING AND INDUSTRY ORIENTATION OF A MULTIMEDIA ENTERPRISE

More than any other business firm, multimedia companies must plan ahead to identify opportunities for developing a competitive advantage in their fields using *technology scouts*. These are sophisticated scientists who can search the world for new ideas and help at every stage of the research, production, marketing and distribution process.

Technology scouts can target multimedia business perspectives in an interdisciplinary manner with other experts. They can find opportunities to form partnerships with other firms, identify barriers to the marketing of new products and lay the groundwork for an intensive commercialization. The information they provide can help senior managers shape corporate strategy.

Multimedia is not the only business where the concept of technology scouts can bear fruit. Such a process has already made its mark in other fields. AT&T and Corning used this approach which led to their joining forces to lay the foundations for the optical fibre industry.

- Corning commercialized the optical fibres,
- While AT&T developed lasers to send signals along them.

Since interdisciplinary research was at a premium, AT&T scientists helped Corning solve problems connected with the process for making the fibres. Management correctly recognized that both corporations would profit from this exchange of knowledge.

The lesson is that the complexity of new products and new markets makes it unwise for companies to try to go it alone in developing and commercializing new business ventures. But loose agreements, as practiced by some industries, are not effective. They tend to be:

- Too slow,
- Too expensive, and
- Too unlikely to produce profitable results.

Hence the spate of mergers and acquisitions in the multimedia industry of which we spoke in Chapter 1 and in this Chapter. The principle is that new products should also be used to further the marketing perspective – and new market vision should be employed in projecting new products.

For instance, Microsoft is counting on Windows 95 to be a Trojan Horse for attacking new markets. The most obvious is the market for online services, using the Microsoft Network (MSN) as infrastructure. Microsoft also has a deal with NBC that relates to interactive multimedia products as well as other plans concerning news-type information on its network.

What may Microsoft do next? Bill Gates described the company's plans to expand further into communications. He said that he also expects to lure more corporate customers, using links between PCs running Windows 95 and server computers using the Windows NT operating system. Along the way, Gates looks to partnerships in entertainment with Hollywood's DreamWorks and, possibly, Turner Broadcasting.

Both in Microsoft's and in AT&T's case, the convergence of skills can be instrumental in creating new markets. As telecommunications, computers and software converge, the leading companies will provide both information and media programming. As a result, the new conglomerates will be both:

- Information superproviders, and
- Content providers.

As we saw in the previous section, the real challenge facing the multimedia companies today and in the coming years is not technology. It is the way to define coming customer requirements and capitalize on them before the competitors do so. This is true for all information brokers and it is going to require well-defined parameters which need to be modelled and manipulated in realtime.

In conclusion, one of the significant aftermaths of the multimedia business is the convergence of industrial sectors. Convergent technologies have a significant economic and social role, but they also bring into the picture legal aspects, as we will see in the next section.

BUSINESS AND LEGAL CHARACTERISTICS OF MULTIMEDIA COMMUNICATIONS

In a highly competitive environment, any global business operation increasingly requires fast, reliable communication among employees and business partners – both clients and suppliers. This must take place in multiple and sometimes remote locations, including realtime visual exchange to accomplish the objectives of an effective business meeting.

Proof of this statement is the fact that *teleconferencing*, which employs state-of-the-art video, audio and computer technologies through broadband global networks, is becoming more widely used in business. It is usually cost-effective, with savings of up to 70 per cent of the travel costs for in-person meetings.

Leading-edge companies now study how to improve upon the effectiveness of teleconferencing solutions using virtual roundtables, including distributed deductive multimedia databases – and virtual reality tools (see also Chapter 7). The motor behind this approach is that it adds value. Benefits include:

- A better ability to communicate, all the way to quicker decision-making.
- Reduced time away from the office, customers and staff, and
- More frequent face-to-face contact with distant colleagues and clients.

Another major advantage is faster and more uniform dissemination of complex project information to multiple groups. Direct connection to customers' and vendors' video facilities promotes timely information exchange and more effective working relationships.

The cost is reasonable because advances in video and communications technology are making teleconferencing both cheaper and easier. This is one of the reasons why, earlier in this chapter, I insisted on the fact that multimedia opportunities will be greater in business and industry than in the consumer market.

Emerging organizational structures designed to work more efficiently in a global market-place are pushing technology towards the integration of audio, video and computer-based information, sharing applications at desktop. This integration is consistent with the strategy of distributed computing and end-user empowerment.

As people become more comfortable with video, and new applications are developed, leading organizations expect a natural evolution from planned, conference-room-based video presentations to less formal, spontaneous use of video at desktop level – but in a global networked sense.

- By installing software in office-based computer systems and mo-

bile terminals, workers on the move can send and receive multimedia information.
- An important requirement is, however, that of developing the electronic infrastructure for virtual integration.

This means a telecommunications system that lets company employees, their suppliers and customers form *virtual corporations* that span all levels of industry, from manufacturing to banking and retailing. A more extensive discussion on the impact of technology on virtual corporations and virtual offices will be made in Chapter 7.

In a way comparable to what has been said about production and marketing expenses versus R&D, companies and the people behind them must keep well in mind that it costs 10 to 100 times more to get business success from technology than it does to get the technology itself. Hence:

- Projects should not only be graded on the research results they deliver, but also
- It is equally important to emphasize what the research contributes in terms of advanced applications and competitiveness.

Another fundamental requirement is the thorough examination of social and legal issues involved in the implementation and operation of multimedia. Both tend to present some new and potentially costly surprises.

In terms of social aftermaths, most of the multimedia drives are advertisers – and therefore propaganda instruments influencing public opinion in significant ways: 'advertising is the right to choose' – among what the different ads are presenting. Many of the legal issues derive from this simple fact and the polyvalence of the multimedia business.

In a ruling that could change the nature of online information services, in May 1995 Justice Stuart L. Ain of the state Supreme Court in Mineola, New York, cleared the way for a $200 million libel lawsuit against Prodigy Services – by deciding that the online network:

- Was a publisher of information,
- Not just a distributor of third-party news.

Prodigy is a joint venture of International Business Machines and Sears, Roebuck. According to this ruling – in response to a motion filed by Stratton Oakmont, a securities investment firm that sued Prodigy for libel and defamation – the company might be held legally liable for comments posted electronically by its users.

The suit arose from a series of messages posted on Prodigy's 'Money Talk' electronic bulletin board in October 1994 by an unknown Prodigy user. The messages portrayed Stratton Oakmont, which specialized in penny-stock issues, and its president, Daniel Porush, as criminals and characterized the firm's dealing as fraud.

True enough, Stratton Oakmont has been investigated by the Securities and Exchange Commission for reported irregularities and securities violations. But so far there had been no convinction. Justice Ain, however, drew a distinction between Prodigy and other information services that merely act as conduits for unfiltered information, much as a bookstore or library distributes information but does not publish it.

Prodigy uses human monitors and automated systems to remove objectionable material posted to its network, a fact which probably motivated the judge's ruling. Other multimedia information providers will most likely work along a similar frame of reference and the decision at the state Supreme Court in Mineola, New York, is a precedent to be kept well in mind.

FINANCIAL APPLICATIONS OF MULTIMEDIA: FROM RISK MANAGEMENT TO HOME BANKING

Production and consumption have always played a significant role in economic and social activities. But as the importance of information becomes ever greater, it is being recognized that its able management has a strong influence on business development and that the effective use of information is a key to success in industrial life.

This paradigm shift where, for instance, information on money is as important as money itself, is now clearly established in business life – but also infiltrates private life as a result of the emergence of multimedia technology such as two-way cable television. New developments seem to be leading us towards another level of communications capabilities, including:

- The provision of sophisticated databases in an easy-to-use environment,
- The establishment of high-speed any-to-any networks, and
- Realtime simulation capabilities for analysis, experimentation and management control.

Realtime simulation with interactive 3-D color visualization, is the technology often referred to as *virtual reality*, of which we will be speaking to a greater extent in Chapter 7. It is a process which has enormous potential in business, for instance as a risk-management tool.

As I have never tired underlining, the first most important implementation domain of multimedia is in *business*, not in the consumer market. For instance, in banking, analyzing an inventory of securities and running 'what if' simulations such as:

- What would be the effect on the bank's position if interest rates rose 50 basis points?
- What if there is a panic in the futures market and prices drop by 10 per cent?

A 3-D colour graphics presentation produces results of conceptual clarity. The software can display a realtime picture for *Value At Risk*, increasingly used by banks and securities firms to determine the probability of profit and loss, within certain boundaries.

The same technology can help in virtual pits. A realtime 3-D view of the trading floor helps to simplify the way we monitor the entire operation – and this can be done in *realspace* embracing the financial market around the world.

- All this is written in the understanding that crisp presentation of fuzzy financial information is a powerful tool.
- 3-D multimedia presentation can substitute for dozens of 2-D computer screens of densely packed figures needed to describe stock market behavior; an example is shown in Figure 5.4.

Every financial market generates huge amounts of information in ways that managers find hard to digest – if they can understand it in the first place. Hence the interest in seeing an entire portfolio encapsulated as a 3-D moving picture on a single screen.

Figure 5.4 Shares Traded (gridded to bid-ask spread vs time of transaction)*

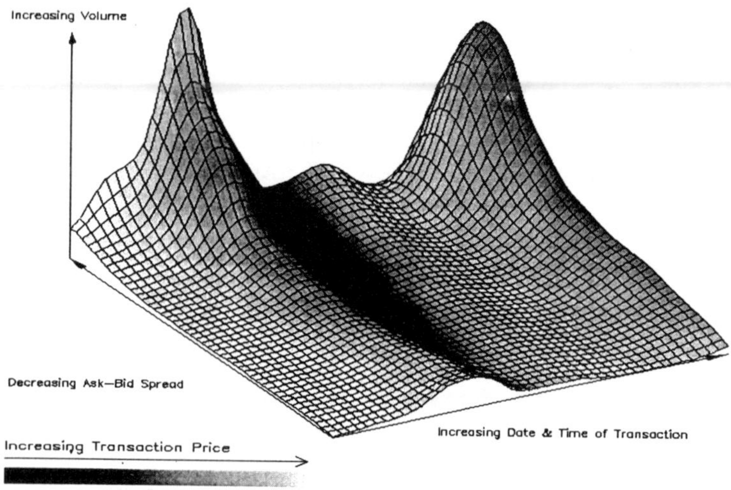

*Plot done with PVWave, published by permission of Visual Numerics

- The image changes in realtime as prices move and derivatives and other securities are bought and sold.
- The user can zoom in for close-ups of particular deals or trading book items, moving freely around the portfolio's landscape.

Tier-1 banks in New York, both investment and commercial institutions, are now using multimedia technology to run a portfolio of mortgage-backed bonds, control foreign-exchange trading and manage their exposure. This is the market on which clear-eyed multimedia vendors will focus as their primary opportunity.

On the other hand, there may be spillovers from the business market to the consumer market. Through *home banking*, for example, it was said that in the *near* future consumers may pay bills, look up their account and generally handle credit and financial matters over the same cable that delivers their TV programmes. This was said in a feature article by Stephen Fox, Associated Press writer, in the finance section of the *Daily Pilot.* The writer then quoted Dr Emily Card as having said that she was afraid the consumer was not ready to deal with all that.

The interview and the article took place not in 1996 but in 1982 (20 June) and Emily Card stated: 'I'm very concerned because I think that consumers will need, in this new electronic age, an information source and perhaps even an advocacy source that is centralized and able to be accessed easily . . . be available over cable and other two-way capacities, whatever way the technology shapes up.'

'Old stuff' you might say, and you would be right. But it is also new stuff because that's the sort of things both banks and multimedia vendors talk about in 1996. In the intervening 14 years, technology has shaped up in polyvalent ways but – with one and only one exception that of ANSER in Tokyo – consumers did not buy home banking. ANSER got a market because:

- Its voice input/output was very simple to use, and
- The cost per transaction was very low – at the time (mid-1980s) the cost was about 5 cents.

Glamorous home banking projects, some based on videotex and others on personal computers, went down the drain. The consumer would not pay their extravagant costs, nor would they bother to learn intricate inputs/outputs. The facts are there. The rest is history.

MULTIMEDIA, VIDEO-ON-DEMAND AND CONSUMER ELECTRONICS

The merging of computers and consumer electronics technology has made market definitions difficult, and the same is true of estimating market potential. Some companies look at themselves as computer makers, others as consumer electronics firms. However both are faced with breaking new ground. They have to move out of saturated markets into new areas – the question is: Which? (See also the discussion in Chapter 9 which looks at consumer electronics from the microprocessor viewpoint.)

One way of defining a consumer electronics product is to say that it is one used in the household. It is typically a product that can end up in the driveway or inside a home, and the bill can show up on a personal credit card.

- This definition, used by many vendors, sounds crisp,
- Technology, however, sees to it that it is rather fuzzy.

Is a personal computer an office equipment or a consumer product? If the PC is left out of the equation who dominates the consumer electronics market? There are also other critical issues – from the protection of intellectual property and the establishment of security guarantees, to bread-and-butter interconnection capabilities.

In all the talk of exciting new communications services such as multimedia, video-on-demand, and interactive TV, one very important topic not being paid appropriate attention is the end-point of the communication channel through which subscribers will access the new services. This is the plain old subscriber loop that links users in to the public network, and this has changed little in 100 years – of which we spoke at length in Part One.

No matter how you choose to look at the subject, access to multimedia networks still largely consists of copper wires buried in the ground or carried on overhead lines. The plant is POTS, therefore old – and as we have seen in Chapter 4 the ISDN talk is for the birds. Yet, the access network has a vital role to play as a service-delivery platform.

- In competitive markets there are new services being launched and investments to be made – such investments needing faster payback.
- There is, as well, widespread interest among network operators in developing access networks that have greater information capacity at lower cost – and work in a flexible manner.

The answer lies in using newer access technologies with optical fibres and radio links, rather than upgrading the existing copper network. A choice of copper, optical fibre, coaxial or radio technologies, with a unified approach to network management, permits the building of flexible access networks.

There are solutions under development today to support all categories of subscribers, from remote isolated houses to apartments in densely-populated metropolitan districts. Market segments range from family homes and home-based workers to multinational corporations – but return on investment in the home market is still a dubious proposition.

The analysis of multimedia business perspectives in the consumer market requires a systematic understanding of what is and what is not affordable, from the consumer's viewpoint, under today's new circumstances. It is not enough to drum up the possibility of 'significant achievements' in the fields of computers and communication technology, software engineering, video-on-demand and other free-market products. We should also extrapolate their impact on consumers.

Even an approach full of creative and instructive ideas will not be a business success unless what is offered is both appealing and affordable. Video-on-demand is often brought as an example of a vast market-in-the-making. I personally doubt that it will be a success, because only a few people are going to pay for it.

- As we will see in Chapter 6, the infrastructure which is necessary for video-on-demand is huge, sophisticated and costly.
- Technically it is feasible to do many things, but it is hard to imagine there will be a rush in a marketing sense.

There may be a niche market for digital movies, and some entrepreneurs try to capitalize on it. For instance, Japan's Masayoshi entered into a joint venture with Nippon Telegraph & Telephone to acquire the rights to digital movies and other software. The goal is to deliver them to 10 million Japanese homes over telephone lines. There is also another similar effort in Japan based on an alliance between NTT and Microsoft (*Business Week*, 1995).

But this will not become a mass market able to sustain all the companies who hope to be major actors – at least not in this century. What *might* become a mass market is telephone line-sharing devices for use in a residential or home office environment, where occasional facsimile usage or basic modem capabilities are required.

With this in mind some vendors, like Command Communications, have introduced user-programmable products designed especially for small offices and homes where automatic answer and follow-me devices, facsimile usage and different modem capabilities are required. By contrast, reheated cold-coffee-type products like 3-D TV vision with toy eyeglasses will be non-starters.

Some market plans are simply day-dreaming. At the Funkausstellung of September 1995, the International Telecommunications and Broadcasting Fair in Berlin, a Japanese electronics manufacturer exhibited its new 2D/3D wide-screen TV.

- Talking from the viewpoint of a prospective user, this product is a total flop.
- Looking at it as a technologist, I find nothing less than trivialities.

Who is the consumer who will pay for video-on-demand on substandard products such as this. No matter if the vendor has used its Modified-Time-Difference algorithm to convert 2-D pictures into 3-D images, the plastic goggles remind of the 1940s, not of the 1990s and beyond.

New circuitry may increase the separation and decrease the noise in the set, and a new wave speaker may also be helpful – but in a marketing sense the end result is trivial. It's a good thing that this silly 3-D television can be watched only by 2 people at the same time, so that there are not so many simultaneously irritated viewers. These 3-D images are a disgrace, inducing consumers to save money rather than spend.

PROBLEMS AND OPPORTUNITIES WITH HYPERMEDIA

The chicken-and-egg problem with consumer-oriented hypermedia and multimedia is that as long as the market is small the prices will stay high. And as long as the prices stay high there is no mass market. This means that there is neither money available for imaginative new solutions, nor can the prices drop – hence, the consumer will not buy. The basics are simple,

- Hypermedia solutions developed as a framework for organizing, structuring and searching information.
- Hypermedia software addressing data models able to implement networks for interconnected information.

This makes sense in business where return on investment rather than absolute cost figures is the decision guideline. Back in 1953,

Dr Neil Jacoby and Dr Louis Sorel, my professors of financial strategy at UCLA, impressed on their students that a banker will be happy to spend a dollar – if he can make two dollars in return. Consumers do not always react in this way, but neither are they totally irrational.

- Several aspects of the technical background for the business market and the consumer market may be similar,
- But the financial evaluation algorithm is not. The application conditions the return on investment.

As a methodology for storing and browsing multimedia information, the hypermedia paradigm works through a system of nodes representing associative relationships. Retrieval is achieved by freehand browsing. Technically speaking, another hypermedia approach is the use of process models concerned with the interactions that occur between a user and the database hypermedia system, including the semantics of linking and browsing.

Sophisticated solutions address various aspects of structure and behaviour characterizing the dynamics and interactional aspects in runtime. Such considerations became critical as, over time, hypermedia systems have become increasingly larger, more distributed and more dynamic.

This evolution makes it increasingly difficult to create and maintain the entire node-link paradigm and, here again, corporate usage has a definite advantage over the consumer market. Whether in a network or in a single computer, navigable links must exist between primitive nodes and they must be global in the sense that:

- They belong to the whole hypermedia database, and
- Can have no local semantic context which is not a global context.

Let us keep these references in perspective when, in Chapter 6, we talk about multimedia servers. The problem with the early hypermedia solutions is that such links are static: they are persistent and exist independently of a user session – in spite of the fact that user sessions get increasingly interactive and dynamic.

Furthermore, with most available software, browsing is two-dimensional in the sense that any desired node can be accessed from a particular starting node by means of link-based browsing, but in a planar manner. A special access approach will evidently be an improvement, but it costs money – from R&D to its implementation.

Not everything of course is bleak. To its credit the current hypermedia solution has the characteristic of simplicity. It is nevertheless no less true that as multimedia applications and their databases expand, the simplicity of the node-link paradigm results in a situation where non-trivial link-based approaches can become lost in *hyperspace* – therefore inhibiting free browsing. Also link editing is tedious.

Other possible shortcomings are that deletion of nodes can cause serious consistency problems eventually leading to mixed semantics when multiple authors create links to and from the same nodes. This results from the fact that links do not generally carry explicit semantic meaning. These problems can affect multimedia approaches on different levels:

- User interfaces, and
- Semantic data modelling.

To solve the problem at interface level it has been proposed to use graphical overview diagrams, link filtering and ranking – as well as the employment of knowledge *agents*. We have already spoken of the fact that knowledge robots will dominate multimedia applications.

For semantic data modelling purposes, the approach is that of defining *metastructures* through knowledge engineering and object orientation. Some studies have addressed logical data modelling by defining common abstractions to combine large amounts of raw multimedia information into a logically-integrated hypermedia database.

As these references help to document, the software infrastructure for the able management of multimedia has to be quite sophisticated. Once again, corporate users are more likely to pay for such supports than the consumers. There is also the issue of new applications. In the next section, we will take as an example the *electronic journal* project, but Chapters 8 and 9 will focus much more thoroughly on the new applications issue.

THE ELECTRONIC JOURNAL PROJECT AT MIT AND NEWS IN THE FUTURE

Under a grant of $150 000 from the Andrew W. Mellon Foundation for use over a period of three years, MIT Press is developing the *Chicago Journal of Theoretical Computer Science* (CJTCS). The goal is that of establishing a model and method for moving forward with electronic journal projects in response to the needs of universities, researchers and libraries.

CJTCS is designed as a cost-based, instead of market-priced, multimedia electronic journal and as a way to introduce more diversified competition into the marketplace. The project aims to:

- Provide a new way of applying cost-sharing and risk reduction to academic publishing, and
- Serve as a platform for experiments in the Internet in connection to publishing.

To fulfill its purpose, the project is using a target community of authors and readers accustomed to daily use of the Internet, and having a convenient access to laser printers. This is a good example of a research effort which addresses professionals and, if successful, can be carried on to the consumer market.

MIT's electronic journal project fits well with projections made by cognizant designers of multimedia systems who see four broad ranges of forthcoming interactive applications:

1. Scientists and engineers who need multimedia to interpret reams of data more efficiently.
2. Business executives who would be able to incorporate video presentations into their computer-composed reports.
3. Students who would use computers connected to optical disks laden with detailed graphics, animation and charts.
4. Home users who would have access to talking newspapers, interactive encyclopedias, books and other repositories of knowledge.

Notice, however, that in the fourth class a major point is made about the use of compact disks and other input devices – not just the online multimedia servers of which we will speak in Chapter 6.

Most importantly, it is the opinion of cognizant designers of multimedia systems that low-cost technology will propel the market in classes 3 and 4. Applications perspectives should capitalize on the fact that:

- Desktop personal computing now merges traditional text and graphics with animation, sound and video.
- Solutions should therefore present a breakthrough in *affordable technology* in order to create new applications domains.

By contrast, in classes 1 and 2 the criterion of success is technological leadership; for instance *layered image representation* which is another of the imaginative projects at MIT's Media Laboratory.

- With layered image representation, image sequences are decomposed into a set of layers ordered in depth.
- They are enriched with associated maps defining their motions, opacities and intensities.

This strategy makes it feasible to synthesize the image sequence using available techniques of *warping* and *compositing*. The challenge is to achieve the description starting with an image sequence from a natural scene.

- Rendering is easy (see Chorafas and Steinmann, 1995),
- But vision is difficult.

Solutions must account for the fact that image coding contains a model as well as an implicit vocabulary. If motion compensation is used for moving images, then regions in the image model move coherently across time. In its fundamentals, the design of an image coding system has three distinct but interrelated processes:

- Encoding,
- Representation, and
- Decoding.

The choice of representation determines the nature of the solution, while encoding converts the representation into an array of pixels for the display. On these notions lies the concept of image layering, capturing important aspects of images and improving the efficiency of image coding as well as the handling of sequences. This approach also permits high-quality frame-rate conversion.

With an image sequence represented as layers, it is possible to perform flexible editing and post-production operations that would not be feasible with classical solutions, where the contents of one frame are estimated by applying a spatial transformation to the contents of a previous frame.

By contrast, in a model based on image layers the image consists of a set of levels with some ordering in depth. Such representations are used by animators, who paint a set of images on transparent plastic sheets as a way to produce the final images: painting a background once and moving the characters in front of it. Research at MIT investigates:

- Perceptual issues in synthesizing a coherent scene from disparate parts,
- Technical issues in the integration of live and processed video, and
- Social issues in the visual depiction of a community.

A Meta-Media project integrates a rich set of graphic tools and editors with searching, browsing, linking, scripting and visualization capabilities to allow research into new design perspectives emerging from realtime, multilayered information in an electronic communications environment. Informational multimedia pathways present a graphical design complexity which cannot be faced through old methods.

6 Multimedia Servers, Communications Solutions and their Software

INTRODUCTION

Chapter 1 and Chapter 5 have given plenty of practical examples to document why, in terms of business and of technology, we are seeing the beginnings of a convergence of computer networking and broadcasting. The business opportunity seems to be there but there is also a host of problems – one of them being skills.

- Few television engineers understand networking, expecially the notions of databases, shared media and interactivity.
- Then again, network engineers don't know much about TV, interactive visualization and multimedia.

This lack of generalized skills able to answer the requirements of convergence is a problem as well as a challenge to be faced by the different technologists. For the time being, much of advanced interactive television issues centre on display and resolution but, as we will see in this chapter, the big nut to crack is *interactive networking*:

- From links,
- To switches, and
- Multimedia servers.

The able handling of all three subjects requires a different culture than that either television engineers, database architects or network designers have had. Convergence brings perspective to the debate over image structure and display parameters. Factual and documented experimentation with solutions is required, the problem being that the call for experimenters may not be familiar either with television or with networking technology.

Here is a brief example of basic differences between computers and television. Existing TV protocols, as for instance NTSC, the American standard, use interlace. That is, half the lines on the screen are updated at a time. By contrast, computers use progressive scan. The difference is not just technical.

- *If* the Information Superhighway goal is to conserve spectrum and see it used effectively.
- *Then* an important issue is to determine which method maximizes the amount of information through the network for a given amount of capacity.

Besides this, digital solutions bring with them a significant change from analog television. The evolving multimedia networks will have in the beginning limited buffers to absorb communications jitter and provide for error control. However, interfaces incorporated into a *set-top box* will rather rapidly evolve towards supporting intelligent agents and other software artefacts to ease the transition.

Finally, some significant developments are expected to take place in the design and use of *multimedia servers*. This is the theme to which a major part of the present chapter is dedicated.

WHO NEEDS MULTIMEDIA SERVERS?

A growing number of computer manufacturers, as well as many start-ups, are investing a lot in media servers and the associated services collectively known as video-on-demand, already discussed in the previous chapter. Because of targeting core utility requirements in the Information Superhighway, it is believed that this market will develop fairly fast, but it will also require a distinct culture and interdisciplinary capabilities.

As the applications perspectives are unfolding, companies which can read the writing on the wall target a dual market perspective, with the reasonable expectancy that many of the concepts and most of the software will be portable between the two client populations which we discussed in Chapter 5: the basic corporations and the consumers. As Figure 6.1 shows, this dual market can be seen in the following manner:

Figure 6.1 The multimedia server will be a high-performance parallel computer addressing two client populations

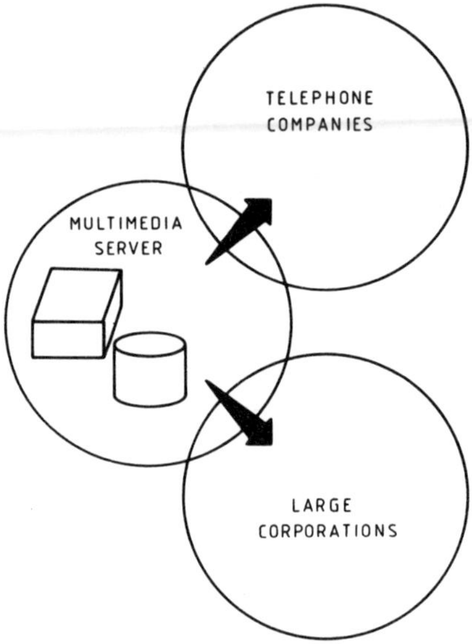

1. The general public interfaced through telephone companies, who are the partners of computer manufacturers in developing the media servers.
2. Large corporations with experience in imaging, which are definitely moving towards multimedia applications well beyond teleconferencing.

IBM is one of the companies building set-top boxes and video servers, but it does not have the culture to make multimedia systems work. Multimedia servers require interdisciplinary developments which are from a totally different planet than that of mainframes, SNA and naive 3270 protocols that the computer vendor is still pushing to its client base.

Cognizant executives at Silicon Valley don't give IBM a chance in terms of recycling itself into the new, fast-moving and highly competitive multimedia arena, even if it buys some start-ups. Several voices have suggested that for companies which have misguided themselves into believing that the world owes them a good living:

- The window of opportunity is getting narrower,
- The upside is getting lower, and
- Room for manoeuver is getting less.

As plenty of systems experts underlined, in their judgment IBM and the other mainframers are still on the wrong track. One of the examples is that IBM is now retraining some 30 000 mainframe-oriented sales representatives to act as 'information superhighway consultants'. But these sales representatives are too embedded into the old culture to be able to create advanced distributed networks.

Among the start-ups, one of the competitors in multimedia servers is USA Video, a company which has some bona fide agreements. It has signed up with partners to manufacture multimedia gear which uses proprietary software. Digital Equipment will build the video servers; Samsung Electronics will make the set-top boxes.

On the implementation side, Rochester Telephone, which has invested money in USA Video, is testing its system in 100 apartments in Brighton, NY. But there may be software and hardware problems. Critics say that USA Video can not digitally compress pictures in realtime, which means it is not able to offer live events, such as sports-on-demand (*Business Week*, 1994).

As these references help to document, multimedia servers are not just another name for classical databases with some data-compression routines. Realtime solutions are crucial not only for corporate video applications, but also for other multimedia implementations requiring an interdisciplinary approach when:

- The line as we know it from WAN and LAN will disappear, and
- Multimedia links and database servers will dominate the landscape.

As far as the links are concerned, we can reasonably expect that we will have to deal with a variety of physical media. For instance: coaxial cable, optical fibres, terrestrial radio and satellite transmission. They are all part of the delivery mechanism and though some ideas exist about which might dominate, this does not change the fact that over many years they will have to interoperate.

Less heterogeneity may exist at the data-link layer where the

Figure 6.2 Technology breakthroughs and business opportunity in the multimedia market

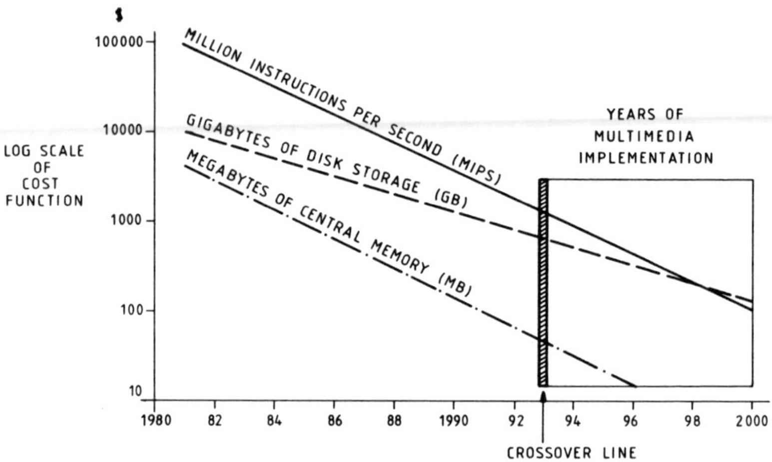

asynchronous transfer mode (ATM) seems to be the better bet (see Chapters 2 and 3). No other network protocol (X.25, SMDS or frame-relay) can provide voice, video, graphics, text and data integration, while frame relay and SMDS might be incorporated as ATM application sublayer definitions within ATM.

Agents will reside at workstations and set-top boxes interfacing to TV, phone and other outlets. Among the software companies, Oracle seems to be at this moment quite alert to the requirements posed by the multimedia servers market. Microsoft and Novel are also positioning themselves to be key competitors.

As these references help to document, technology underpins the business opportunity which is developing in the multimedia market. Figure 6.2 demonstrates that it is no accident that 1993 has been the year when interest in multimedia servers became concrete and solutions have been actively sought after:

- Increasing cost-effectiveness has created a *crossover line* where the implementation of multimedia becomes a practical proposition at affordable cost.
- While the unit cost per million instructions per second, gigabyte

of optical disk storage and megabyte of central memory will continue dropping, current levels are low enough to permit practical multimedia applications.

Aware that the years till the end of this decade will be crucial in establishing a leadership in this domain, telephone companies are moving in to capitalize on the experience of software and hardware manufacturers. We have spoken of this trend in Chapter 5 and said that it is expected to continue and gain momentum.

Some financial analysts, however, wonder whether the telephone companies with their bureaucracy and consumer orientation are the best positioned to develop and market multimedia servers. Let me repeat for the nth time that those who will need the most multimedia servers as the virtual company takes hold (Chorafas and Steinmann, 1995), are financial institutions and industrial corporations. As for the mass market the criterion is very low cost – and this is not obtainable with the solutions that old-fashioned telcos have in their minds.

CLIENT-DRIVEN, SERVER-DRIVEN AND COST-EFFECTIVE APPLICATIONS

Telephone companies who think they may dominate the multimedia server market may be in for a rude awakening, not because of want in investments but because of differences in culture. Bell Atlantic has with Oracle and nCube a $50 million contract for the development of multimedia servers. British Telecom has acquired three nCubes and, in collaboration with the same two vendors is working diligently in developing solutions.

But as cognizant executives and systems experts were to underline during the research meetings which I had, the effective development and implementation of media servers requires a very significant cultural change. This must go beyond the transition from data to multimedia, changing the systems design:

- From *client-driven* applications, where the environment is simpler and less demanding,
- To *server-driven* applications, which can be quite complex and require sophisticated software.

The digitization of video content is in itself a significant task but it is not an overwhelming development. As we will see in this section, there are many factors which come into play creating among themselves a new technological landscape, with the conversion from client-driven to fully-distributed server-driven applications being at the pivot point of the transition.

The change in culture rests on a number of technological breakthroughs which have progressed in parallel, but few people properly correlated them in a way to develop a pattern. Yet, the synergy from these developments revolutionizes what we know so far in terms of computers, communications and software. The five top references are:

1. Massive infrastructure improvements in an information superhighway sense;
2. The merger of computer technologies with HDTV and digital audio;
3. The move towards realtime encoding of information in a digital form, with growing volumes available digitally;
4. The availability of efficient compression technologies; and
5. A new generation of computer-literate users in the First World.

Taken together, over and above the sharp drop in the price of MIPS, gigabytes and megastreams, these five reasons see to it that a sort of *digital demand* has arrived and it is pushing for new media servers – but not at any price. Market demand increasingly calls for cost-effective solutions – even if under current technology the $100 000 media server supported through a parallel computer seems to be set as a target.

If this level of cost and performance focuses at the distributed nodes of the multimedia environment, upstream at the main nodes are projected solutions involving a multimedia server with 1024 processors, 16 gigabytes of central memory, 1.5 terabytes of disk memory and 512 I/O (input/output) channels (128 for input, 384 for output) able to support:

- 1500 title video databases, and
- Over 10 000 simultaneous video streams.

Figure 6.3 exemplifies the solution adopted by Oracle and nCube,

Figure 6.3. Multimedia architecture advanced by Oracle and nCube

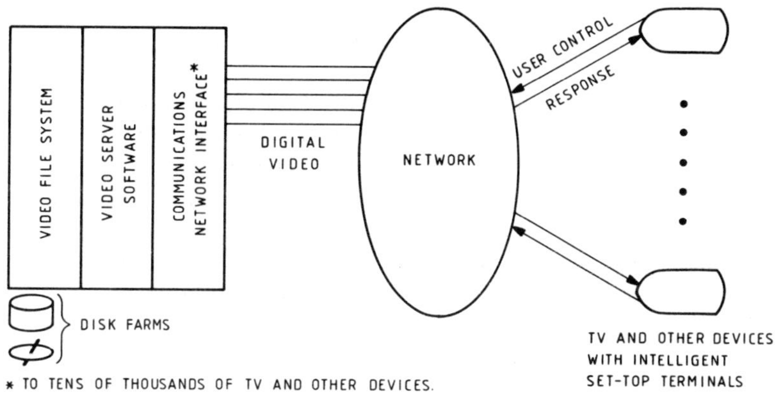

on the basis of its collaboration with Bell Atlantic and British Telecom. It is a multimedia architecture supporting both magnetic and optical-disk farms, a video file system, video server software and communications network software addressing digital video.

Bell Atlantic has asked for large disk capacity able to handle thousands of titles. It takes 1 GB of magnetic disk storage for 90 minutes of video, under current compression ratios. Another crucial requirement is high I/O bandwidth. The developing norm is about 200 KBPS for each video output. This is equivalent to a T1 line – but ongoing research is targeting beyond 1.5 MBPS towards 6 MBPS channels.

Still another basic requirement regards large switching capacity. Multimedia information from many disks may be connected to any output channel. Since resources are not unlimited, sharing must be done to permit prices to be attained which the market can accept. This calls for high-grade switching capacity which can be steadily improved. Other needs include:

- Realtime execution at subsecond speeds,
- Dramatic scalability in video streams,
- Reliability with at least 99.9 per cent uptime,
- Massive capabilities to handle thousands of titles, and
- Versatility reaching VCR-like capabilities.

The crucial issue of a general architecture incorporates all of these characteristics and supports video pumps which work fully in realtime. But will the cost be reasonable? Current cost targets stand at below $600 per stream, and I personally think this is way too high for the consumer market.

Cost criteria for the business and industrial market are a different matter, and the opportunities are there if the different vendors know how to identify them and exploit them. Technological and business perspectives integrate in different ways at the home and in the office:

- At the home market, the telcos plan to use set-top boxes and automate access to the video pumps through credit card numbers.
- In the office, the crucial criteria for success are quite different – as we saw in Chapter 5.

While the success of consumer-oriented business propositions lies in catching public attention – to avoid another painful experience like that of videotex – there is a great deal that can be done on the business side of the information superhighway without false promises and hype. An example is the use of multimedia servers to underpin a corporate memory facility (CMF).

- The best reference to a current application along the CMF line is *Magellan* by Bankers Trust.
- Other business opportunities include 24-hour banking and global risk management.

The potentially lucrative implementation areas range from the trading environment to the realtime calculation of the aftermaths of risk and exposure. Major financial institutions today see practically half their yearly profits disappear in writing-off bad loans. Even a 10–20 per cent improvement in quality control through interactive multimedia may bring about a torrent of investments.

Oracle and nCube see another business target for a parallel computer architecture once the software has been developed; that is a telecom *multimedia switch*. This concerns the private branch exchange (PBX) market, where the new generation of multimedia engines may provide valid answers – but computer vendors like Datapoint and Nixdorf who ventured into new PBX failed. This is still a market controlled by the manufacturers of telecommunications gear.

Figure 6.4 Multimedia server technology leads to the merger of formerly distinct entities into new types of supports

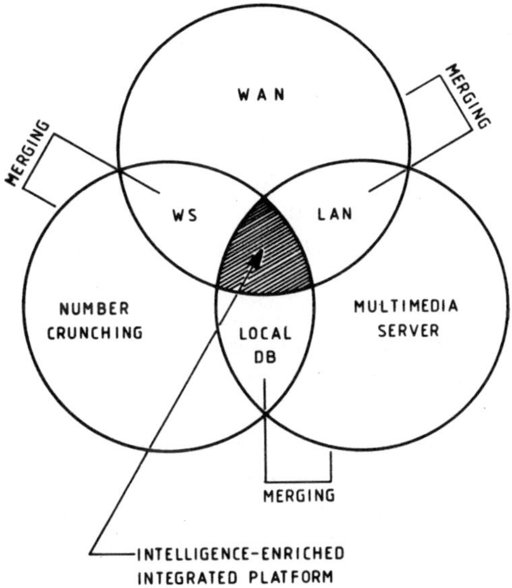

DEVELOPING SOPHISTICATED SOFTWARE FOR MEDIA SERVERS

As we have seen in the previous section, one of the characteristics of the evolving computers, communications and software landscape is the transition from a client-centred to a distributed server-centred environment. But there is also another axis of reference in the new landscape and this is the merger of technologies, which is demonstrated in Figure 6.4. It combines into one aggregate:

- Wide area and local area networks,
- Workstations and number crunchers,
- Multimedia servers and local databases.

At the centre point of this graph is the provision of an integrated platform for intelligent software supporting the user-actuated implementation areas. The new solution call for sophisticated programming modules all the way from the handling of compound electronic documents to demand site management.

Most evidently, such demand cannot be met without an agile interactive software development environment which starts at the level of the user's workstation and ends at the network connection. With desk area networks (DAN) coming into the picture, this will soon mean parallel software development perspectives.

An agile systems design should address all of the tools, assuring that applications are developed at workstation level and run effectively on the network, its nodes, multimedia servers and parallel computers. The strategy which is now shaping up can be expressed in this way:

- Develop on the workstation(s), and
- Run on the network nodes.

This brings up the question of parallel programming efforts, where both the tools and the skills are still wanting. In Berlin, GMD-FIRST (the parallel-computer project sponsored by the German government) parallelizes a recently developed serial program through the assistance of learning genetic algorithms. This is indeed a possibility provided one does not try to use genetic algorithms to parallelize old programs.

At Silicon Valley, nCube has chosen a different approach. The compiler of the *parallel software environment* (PSE) it has developed will do parallelization through standard codes, rather than changes in the syntax. In company jargon, the term 'standard codes' refers to standard parallel communications libraries.

- The end-user or systems programmer would specify the number of professors to be allocated.
- Knowledge-enriched software will permit working with a variable number of processors, giving feedback on parallelization.

Performance-analysis tools, provided by the vendor, can assist in this feedback, including explanations on reasons for variance. For instance, for a prevailing load imbalance or other issues affecting system performance.

Associated to this effort is an end-user visual programming environment permitting the management of interactive video communications under Unix. The user will not need to address each processor. Routing will be done automatically to the right video pump.

As the reader will appreciate, there is in evolution a new software development methodology focusing on the management of multimedia servers, looking at the latter not only as a *device* but also as *processes*. This is in line with the new philosophy in management practice which suggests that:

- It is more important to *focus on an issue*,
- Then the outcome of that issue.

By not focusing on the processes characterizing the user-oriented application, we risk destroying the project which we have undertaken by diminishing its accuracy and its appeals. Quite similarly, by not focusing on the subjects on which it deals, management risks destroying the whole firm.

An example on focusing in the case of multimedia servers can be given by bringing under perspective the *layered images* of which we spoke in Chapter 5, in connection to a project done at MIT. The way this project looks on image analysis, each layer has its own:

- Motion,
- Depth,
- Pattern of color,
- Intensity, and
- Pattern of transparency.

The information stored in a layer can extend over a larger visual angle than is presented on a given image. When data is accumulated over frames the model considers how the individual samples in the frames are to be mapped into the accumulated layer. The latter has a sampling lattice, and the incoming images generally present samples that may not align precisely with those of the layer.

In cases, different images in a sequence may have different degrees of resolution. Then, a Laplacian pyramid is built to combine the information from each image (see for example Chorafas, 1996), and the pyramid coefficients from the different images are compared. The coefficient with the largest deviation from zero is retained.

By itself the layered representation does not necessarily offer data compression. To achieve compression, representation is used in conjunction with other image-coding techniques. Each of the maps in the layers can be compressed using transform coding or sub-band coding.

Focusing permits analysis of a scene into layered representations, and this requires grouping the points in the image into multiple regions where each region undergoes a smooth motion. The motion model used in the analysis helps to determine the quality of representation.

As these examples help to document, the able handling of multimedia requires a different approach to software development than the one required by legacy systems. No better case can be presented to further document this statement than the process of semantic image modelling which is explained in the next section.

MULTIMEDIA SOLUTIONS AND SEMANTIC IMAGE MODELLING

The MIT researchers have chosen an affine-motion model to describe a wide range of motions commonly encountered in image sequences – which include translation, rotation, zoom and shear. Typical methods in multiple affine-motion estimation use an iterative approach to detect multiple-motion regions in the scene. The analysis of an image sequence into layers consists of three stages:

- Local-motion estimation,
- Motion-based segmentation, and
- Object-image recovery.

The motion-segmentation algorithm is divided into local-motion estimation and affine-motion segmentation. The images of the corresponding regions in the different frames differ only by an affine transformation.

By applying these transformations to all the frames, the Media Lab project aligns the corresponding regions in the different frames. With motion parameters accurately estimated, objects appear stationary in the motion compensated sequence. However, some of the images in the compensated sequence may not contain a complete representative set.

To recover the complete representative image and boundary of the object, the MIT researchers collect the data available at each point in the layer and apply a median operation. This can be seen as temporal median filtering on the motion-compensated sequence, in regions defined by the region masks.

- This context makes the interpretation not only easier to comprehend but also flexible in terms of mapping.
- It permits a two-way interaction between available contextual information and the models used to represent multimedia information.

The ultimate goal of the project is that of putting semantic meaning into information theory. Effective execution of semantic meaning requires realtime modelling, which becomes feasible as computers begin to approach gigaflop speeds at affordable costs – and data networks more towards gigastream bandwidth.

With such technological support, it becomes possible to interact in realtime with fairly complex quantitative and qualitative models. In an information-rich environment where data, images and sound are readily accessible and digitally communicated, the need is felt for:

- Semantic representation, and
- Content-based search.

One of the projects at MIT is developing a repertoire of graphics that will allow computational assistance in the expression of dynamic and interactive design (see also an application example on multimedia digital preassembly in the following section). An electronic information environment needs new graphical principles, tools and editors – which are suitable to the interactive, dynamic and intelligent presentation of information.

In this and other similar projects, graphical sets are being integrated with realtime design-assistance systems in order to cope with the magnitude of visual complexity resulting from multiple streams and forms of data that deluge the user. Breakthroughs are necessary that bridge the gap between:

- The hands-on world of designers, and
- The more abstract symbolic world of programming.

The goal is to be able to explore spatial, temporal and relational rules and methods which rank information for the viewer and influence his responses. Multimedia servers designed for professional applications should fully account for these facts.

Depending on the application, realtime semantic image modelling may imply an extremely demanding environment that requires the integration of multiple degrees of freedom of input/output with state-of-the-art computational processing. Effective technological interfaces must sense user-activity on a wide range of length scales, ranging:

- From less than a millimetre, in case of stylus input,
- Through centimetres for gesture sensing,
- To metres for local tracking, and
- To kilometres for navigation.

The scope of multimedia servers is bound to expand as, increasingly, data input and visualization requirements must be sensitive to this wide range of length scales. Sophisticated applications see to it that these measurements must be:

- Done in 3 dimensions,
- Produce images as well as metrics, and
- Maintain the required spatial and temporal resolution.

At the output end, there is need for the development of intelligent graphical tools and principles to support the interactive creation of symbolic information and its landscapes. This must be supplemented by the integration of such landscapes with pictorially convincing virtual environments, enabling a multimodal natural language communication with the virtual environment display and its contents – through combinations of multimedia output channels.

THE USE OF MULTIMEDIA IN A PROCESS OF DIGITAL PRE-ASSEMBLY AND REALTIME SIMULATION

The rigorous solutions which should be developed for compound electronic documents and other multimedia applications must assure that both *flexibility* and *visibility* are enhanced. Visibility is improved by having all product information logically linked and provided to users, automatically for any given task.

A different way of making this statement is that the design and implementation of multimedia servers must take full account of the applications to which they will be put. I particularly underline this issue because somehow:

- Most of the work done today on multimedia servers targets the consumer market, and
- It is conditioned by the hype that there will be a great market for video-on-demand.

Companies working along this line of reasoning will be in for rude surprises. By contrast to what is currently 'en vogue', research on multimedia servers should be oriented to professionals and include polyvalent data management, network services of high reliability, agile software, as well as information modelling and analysis.

It is top management orientation, rather than the technological fundamentals, which I find to be wanting. One of the central technology themes in this process is *digital pre-assembly* (DPA), whereby all of the individual modems are brought together to:

- Detect clashes between parts, and
- Allow engineers to view their designs in context to one another.

Known from the simulation of complex engineering systems, this modelling approach has supplanted the traditional physical mock-ups that were costly and time-consuming to build. Digital pre-assembly solutions have led to a significant evolution of existing process automation. Examples are:

- Assembly modelling,
- Concurrent assembly of mock-up(s),
- Total product modelling,
- Virtual prototyping, and
- Virtual manufacturing.

These multimedia applications rest on *realtime simulation* and appear to have enormous potential for increasing competitiveness in manufacturing and construction. But they also need a careful examination of multimedia server support, and better definitions of systems and procedures – to be properly integrated into an effective strategy.

It is appropriate to remember that as simulation and *virtual prototypes* replace physical prototypes, and these approaches become the norm in product development, not only full understanding of the simulation goals but also thorough testing of the simulator itself become a necessity.

- A sound policy is that of checking results and looking for errors, on the assumptions that there *will* be discrepancies.
- Don't ever listen to arguments that even if some errors exist they will have little or probably zero effect on most calculations.

Along the same line of reasoning, the definition of multimedia server characteristics will not be static. Recent advances in technology may be changing what was previously a niche market into a mainstream information channel. By now it is not that difficult to predict the long-term effects of exploiting multimedia capabilities, because progress in business and industry applications highlights their potential.

Leading-edge user and vendor organizations are talking in terms of digital pre-assembly – and therefore of multimedia systems – rapidly becoming the principal implementation in tomorrow's information technology. However, in a practical sense a number of queries still remain such as:

- What system infrastructure is required by multimedia applications?
- What should be added to the user interface to be correctly implemented?
- What can and cannot be achieved with today's server solutions?

The only way to answer these queries in a factual and documented manner is through hands-on experience in realtime applications. This is why digital pre-assembly is an important project which helps to highlight strengths and weaknesses, as well as in shortening the time-to-market of new products brought to the market-place.

Sophisticated applications help in deciding how to resolve key technical and commercial issues – and this is significantly helped through multimedia systems. They also assist in projecting the benefits to be obtained from advances in technology when new products emerge. Such applications require a coherent plan for utilizing the potential of interactive multimedia to further business objectives in an environment of:

- Rapidly changing technology, and
- Increasing business pressures.

Solutions along the lines of those referred to in the previous section and in this section are further promoted by the proliferation of *bite*, or economy versions of formerly high-priced software. The appropriate utilization of new market offerings, designed to provide better value, must be examined in the most careful way. Such examination should take place in a steady manner because of the need to upgrade system components as the state of the art evolves. This is as true of multimedia servers as it is of network protocols and other artefacts.

MULTIMEDIA SERVERS AND THE ASYNCHRONOUS TRANSFER MODE (ATM)

The whole problem of what constitutes a modern communications network, as well as how to best support connectivity to this network, has to be rethought with multimedia servers. The difference between a call and a server connection is not a mere issue of semantics. It is a very significant technical question requiring precise answers.

On balance is the future relation between the telco-dominated world of circuit switching and the LAN-oriented market in which many corporate multimedia services are being developed. This issue returns to the dichotomoy between a consumer market and a professional market of which we spoke in Chapter 5.

- The current circuit switching, narrowband model for communications couples the call with the connection.
- The connection is the underlying signalling information that reserves a fixed amount of bandwidth and a path.

This definition reflects the fact that a voice call needs 64 kilobits per second (KBPS) in the ISDN world. The network knows this, as well as that the called party is also a voice user with a voice terminal. But such a model does not fit well with multimedia. It tends to be counterproductive.

Because of their bursty nature and other implementation characteristics, multimedia solutions of all types – and most particularly those for business and industry – require an easy addition of bandwidth. Also, a very flexible addition or deletion of applications by users:

- Some end-users may be talking from a desktop PC with video,
- Others may employ desk area networks with multimedia facilities,
- Still others may be using nothing more than videotelephone.

Under today's call model and signalling approach, network call control and connection control are bound together. Therefore, each call would have to be taken down completely and re-established with the new parameters. This is neither an efficient nor an elegant approach with multimedia.

True enough, this subject did not come up last night. The call–connection separation arose in the late 1980s as carriers were grappling with intelligent network call models. But so far it has not been possible to change the call–connection model, largely because of the difficulty of retrofitting plain old telephone services (POTS).

- In the meantime, new challenges arise as multimedia must be complemented by more agile signalling concepts, and
- The prevailing opinion among carriers is that the broadband asynchronous transfer mode (ATM) is the best way to achieve this goal (see the discussion on the technical characteristics of ATM in Part One).

The good news is that every router manufacturer has announced an ATM interface and several ATM switches are already shipped. Also, several users are in the procurement stage of implementation. At the same time, some computer manufacturers have announced ATM interfaces for workstations, while the cable companies recognize the need of running ATM protocol over their coaxial-cable infrastructure.

The bad news is that the detailed technical issues are somewhat confused at the transport level of the ISO/OSI model. Including the case of ATM, data link layer protocols lose packets or cells. Hence, a transport layer remains necessary to impose the appropriate quality of service over end-to-end circuits.

- The Transmission Control Protocol (TCP) *might* be suitable for video applications in which the user is willing to sacrifice full motion to gain error control.
- But for many cases, such as entertainment, this is the wrong balance – and the same is true with certain professional applications.

Today, there are several experimental protocols designed to modify such behavior: GAM-T-10, Delta-T, NETBLT, VMTP, the Internet Engineering Task Force proposals and High-Speed Transport protocol (HSTP). But they are not considered to be products of a broader vision of networking and may not scale up to the gigabit capacities necessary for multimedia implementation.

Higher-up in the ISO/OSI model is the session layer – where quality of service is negotiated between what the underlying communications system can offer and what the application needs. With the advent of data link layer protocols like ATM, and transport protocols, it becomes quite important to redefine what the session layer should support – and how this can be done in the best way.

One experimental setup employs Partially Error Controlled Connections (PECC), a regime that lies over the transport layer and under the application layer – and also includes data compression. The fact, however, remains that the Internet protocol is very primitive compared to what is needed for video.

Finally, at ISO/OSI layers 6 and 7: presentation and applications, the minimum which needs to be supported is electronic messaging and virtual terminal capabilities. The virtual terminal characteristics should not only address bufferless television and video conferencing but also, and most particularly, interfaces through set-top boxes or software running on PCs.

Hence, at the communications end between multimedia servers and workstations, there exists a bifurcation. The upper ISO/OSI layers are in for a thorough redefinition – while at the data link ATM provides a common protocol across local area and wide area networks and can also encapsulate other access protocols. What still has not been settled is:

- How the signalling should be adapted to accommodate call and connection control separately, and
- How should network management systems give precedence to certain calls, unlike in today's environment where every call is treated the same way.

At least one vendor, Bosch Telecom, has a notion known as a *facilities message*. It is a datagram that allows end systems to negotiate a level of service they would like to use across a reserved piece of bandwidth, before bandwidth to the call is allocated. For its part, the International Telecommunication Union (ITU) Standardization Sector is exploring a number of facilities message options, but no action is expected in the next year or two.

PROVIDING THE NECESSARY SYNCHRONIZATION IN MULTIMEDIA APPLICATIONS

Most information technology applications are made in heterogeneous environments (see also Chorafas and Steinmann, 1993). Nowhere is this more true than with multimedia which is par excellence a combination of heterogeneous types of data such as text, tables, digitized images, vector graphics, animation and sound. Among these media:

- Text and images are time-independent,
- Sound, animation and video are time-dependent.

An image or text appearing on a display terminal, for instance, is considered complete as soon as it shows up on this terminal. By contrast, displaying a video clip or playing some voice message is not considered complete until the recorded voice message or the clip reaches its end.

Voice and video interact in some way with time, and their completeness depends on the passing of time. They also vary with time and therefore are called time-varying information elements, while text and data are time-invariant.

- Compound electronic documents typically have both time-dependent and time-independent components.

- Such coexistence usually includes some kind of interaction between the different media, making the solution more complex.

An effective interaction is the kernel of multimedia applications and can be classified as belonging to either of two types: *Sequential* and *Overlapped*. Sequential is an interaction which could display some text describing a given site. Then as the text ends, an animated clip may show that site.

Sequential interactions are rather simple given that the start of any action is activated by the end of another. Therefore, it is not that difficult to create an application applying such interaction as long as its sequential nature is clearly defined and preserved.

Overlapped interactions generally include time-dependent media, though they may also incorporate time-independent media. For instance, an overlapped presentation can be made by displaying a text as soon as an image appears – which is a time-dependent approach. Therefore, simultaneous appearance should be a managed process.

This can be handled successfully if we achieve effective synchronization. Synchronization is necessary to manage many aspects of media interaction, as it assures that actions performed by one system component are in accord with the other system components in regard to application-oriented requirements. It follows that:

- A multimedia architecture should be able to handle synchronization problems which characterize the data flow and media interaction requirements.
- The process of synchronization depends on time and, as such, it presents time constraints which may be independent of the medium involved.

This problem is not necessarily new, but present-day synchronization schemes, provided by multitasking programming languages and transaction-oriented operating systems, are rather primitive in comparison to what is necessary with multimedia. For instance, applying semaphores for resource reservation purposes can be grossly insufficient in a heterogeneous multimedia environment.

The rendezvous synchronization method, which is implemented in some languages, also falls short of requirements when compared with multimedia constraints – in the way we have seen them in the preceding paragraphs. This is particularly true with parallel processing which, as we will see in the following section, constitutes the best solution for multimedia.

There are several reasons for this statement. One of them is that in a single processor environment, a rendezvous taking place between tasks is implicitly given – but the time is not determined and the waiting period of one task for its rendezvous with another one is generally not restricted.

- Parallel computing can correct this situation, but it requires sophisticated synchronization primitives.
- These primitives should allow the designer to specify the time instants when a synchronization is to take place.

As these references help to document, there are plenty of problems to be addressed in a multimedia communications environment. The challenge of optimal server design is not just limited to the device itself, but the choice of a device – and most particularly of agile software – may be instrumental in finding a solution as will be demonstrated in the next section.

THE SEARCH FOR A MULTIMEDIA SOLUTION BY ORACLE AND nCUBE

nCube has been one of the early parallel-computer companies to adopt a hypercube architecture, which is shown in Figure 6.5 (see also Chorafas and Steinmann, 1989). The firm remained steady in this architecture while other firms like Intel Scientific and Thinking Machines changed: Intel Scientific altered its supercomputer design to mesh, and Thinking Machines to fat wire.

Hypercubes provide the sort of parallelism which fits best a multimedia server implementation. nCube capitalizes on this advantage by software contributions provided by its sister company Oracle – which aims to convert its product to the first DBMS able to handle multimedia, including:

- Storage,
- Retrieval, and
- Traffic requirements.

Figure 6.5 A 16-node hypercube architecture

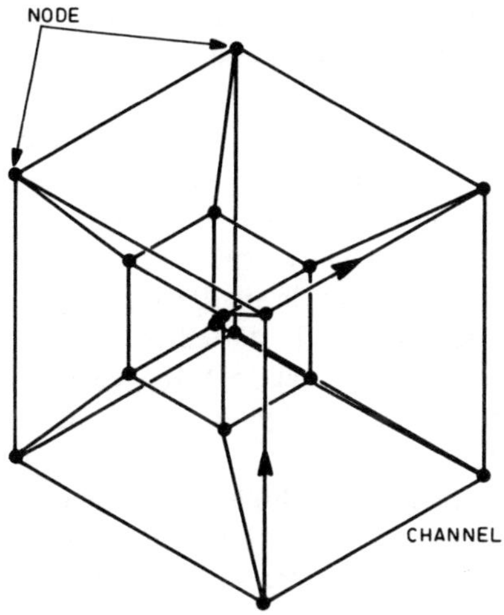

A 16-NODE HYPERCUBE

Support of these functions by multimedia servers and disk farms calls for a strategic level change not only in the way software is designed and built, but also in terms of company orientation and business. As late as 1992, nCube derived 90 per cent of its income from scientific computing while now, in terms of business, scientific applications constitute the minor part; the major share is going to disk farms and to media servers.

Along this line of reference, nCube is targetting three markets. Over the next few years, these are expected to bring $1 billion of annual business to the company, provided management is able to tune its product line towards the following business framework which constitutes nCube's strategic plan:

1. 60–70 per cent of annual income from databases and disk farms,
2. 15–20 per cent of annual income from media servers, and
3. 15–20 per cent of annual income from scientific applications.

In the database computer business, through a close collaboration with Oracle, nCube capitalizes on redundant arrays of inexpensive disks (RAIDS). The next bet is that over the years as the Information Superhighway tunes into operation, income from business line No. 2 will significantly increase.

Other vendors of media servers make a similar projection. Their strategic plans rest on the premise that as parallel technology advances, so will the multimedia server market – which means, however, an evolution in product orientation.

This evolution can be seen in the design of the nCube. The company's line of business, which started in 1983, has moved from a limited number of processors at 15 MIPS each to a new generation (nCube 3) which can feature up to 8000 processors at more than 150 MIPS each.

The chips used in nCube's high-performance computer are custom-designed for reasons of *silicon-level integration*. According to the developers, silicon-level integration provides both:

- Higher reliability, and
- Higher performance.

Evidently, other vendors contest this argument – their thesis being that the development of commodity microprocessors is so rapid that designing custom chips is a waste of time and money. The commercial battle of the late 1990s will be fought in new types of software rather than in hardware – and nowhere is this statement more applicable than in relation to software for multimedia servers.

That parallel processors deliver much more power to the dollar is today beyond question. As Figure 6.6 documents, there is a huge difference between what hypercubes can deliver – versus the classical Vaxes, other maxis and mainframes. In terms of cost-effectiveness in transaction processing, for instance, the ratios between mainframes and high-performance computers are above an order of magnitude.

- Tests focusing on cost feature ratios which average around 1:45
- Tests oriented on effectiveness lead to ratios of about 1:36

Thus, both present an advantage over mainframes well beyond an order of magnitude, but the question of best solutions for multimedia

Figure 6.6 Cost-effectiveness in transaction processing

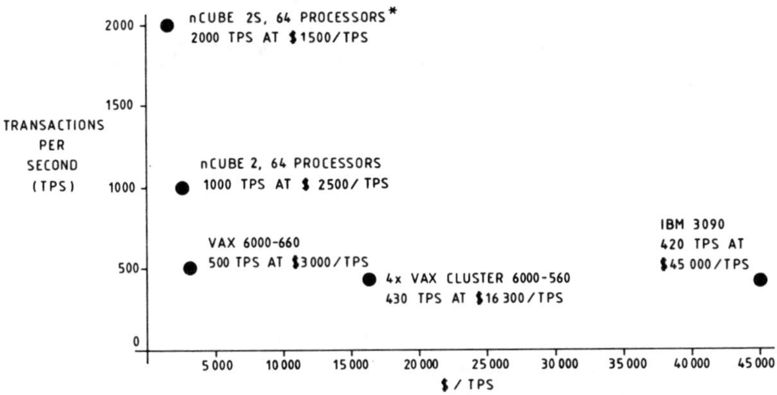

* NEW 512 PROCESSOR VERSION WILL FEATURE 15 000 TPS AT $1 000/TPS

servers is still open, and the same is true about design issues connected to protocols – as we saw earlier in this chapter. Vendors of system solutions for multimedia, as well as telcos, will be well advised to put their act together and show evidence of excellence at an affordable cost. Only then will the market for multimedia applications become a lucrative line of business.

7 Practical Applications of Multimedia and Virtual Reality

INTRODUCTION

To a very significant extent, competitiveness in the mid- to late 1990s will be defined through interactive multimedia services. In its many ramifications multimedia is an emerging market, growing out of the current ferment over broadband channels not only in communications but also in input/output (I/O). This extends all the way to multimedia computing and involves the linkages between:

- Cable television,
- Satellite nodes,
- Telephone lines, and
- Personal computers.

Solutions under development feature the integration of text, data, graphics, sound and video into a single system. Supported facilities will allow both office and home computer owners to use their machines in polyvalent ways: business transactions, learning, entertainment and shopping – leading to 2-way television.

As Chapter 1 has documented, this development underpins a number of recent merger, acquisition and investment activities, all involving multibillion dollar investments. Motivated by market potential, Pacific Telesis and other telcos are bringing fibre-optic cable to offices and homes.

Chapter 1 underlined the need of revamping the human interfaces. Broadband I/O is not obtained just by doing something a little better at workstation level. Neither is the now classical mouse, much less the keyboard, the means to provide bandwidth. This can only be done through practical applications of *virtual reality* (VR).

Virtual-reality technology now stretches well beyond the halls of video arcades; it is being used in industry as a serious engineering tool. Applications areas showing the greatest promise in the next four years are computer-aided design, architecture, realtime simulation and training. Many advanced implementations in finance fall in the simulation domain (see also Chorafas and Steinmann, 1995).

- The practical implementation of virtual reality enables users to navigate through an onscreen, computer-generated 3-D environment.
- In this virtual platform, they can change their viewpoints and manipulate objects which they themselves created, or have been introduced by the market.

Among the benefits of employing such an interactive system is that end-users can manipulate objects onscreen before the actual products are built or commitments made – checking for appealing features, design flaws or contradictions. This chapter examines a number of practical applications which could be significantly enhanced using the services of the information superhighway.

VIRTUAL OFFICES AND THE GROWING NEED FOR INTERACTIVE MULTIMEDIA

The concept underpinning virtual offices can be explained in a few sentences. By installing software in office-based and home-based workstations as well as mobile terminals, workers on the move can send and receive data *as if* they were in their office using conventional computer systems, but at a much higher level of flexibility.

The first prerequisite for such solutions is cultural, including the ability to appreciate the business opportunity which it presents. An equally important requirement is developing the communications infrastructure for virtual integration. That is, an intelligent network that lets:

- Company employees,
- Their suppliers, and
- Their customers

form *virtual corporations* able to span all sectors of industry, from manufacturing to banking and retailing. As we have seen in Chapter 1, merging cable TV and telephone business expertise will allow new technologies to be exploited, such as *interactive media*, for virtual-office purposes.

This is the reason why a software company with a hold in the office market, such as Microsoft, has been in talks with Tele-Communications, Time Warner and other companies about possible alliances to help define, then capture, some of the evolving business opportunities. Microsoft has also stepped up its efforts at developing so-called *content*, in connection to educational and information programs:

- Featuring sound and high-grade graphics,
- That will run on multimedia personal computers.

Are virtual-office practices appealing to their intended users? In principle the answer is: 'Yes', though there is still no rush in this direction. As a study by the German Bundespost helps to document, there are two sorts of responses in connection to office employees, and the reader is advised to take notice of the statistics:

- The large majority, between 92 and 95 per cent, wants to maintain the social environment of being in the office.

The reference to the type of employment includes not only clerical work but also managerial. A very significant fraction of the population does not want, or at least not yet, the global office supported by the information superhighway.

- The minority of between 5 and 8 per cent prefers to work at home, making up its own timetable and largely depending on computers and communications for contacts.

Given this dichotomy, a key problem is the motivation of the working population and its opposite: the avoidance of overexpectations in terms of results. But 5 to 8 per cent as a starting base is not that bad after all. It provides an environment for both real-life technical tests and for a social experiment which, if properly managed, has a good chance to succeed.

In terms of a practical business implementation, the goal is to bring multimedia into the growing world of client–server computing, where powerful servers run large databases and dish-up information and programs to workstations across networks. Hence the race to create software needed to store a variety of applications that will flow through the information superhighway:

- Electronic mail,
- Commercial transactions,
- Financial deals.

Not just newspapers, magazines and entertainment video. However, in all these projects companies and people behind them must keep well in mind that it costs 10 to 100 times more to get business success from technology than it does to get the technology itself.

The advice that this last paragraph conveys is that projects, as well as laboratories working on them, should be graded not only on the research they deliver but also on what the research contributes in terms of:

- Advanced applications, and
- Competitiveness.

The information superhighway should be a practical experience with tangible deliverables. This is a message not yet understood in all quarters, particularly among government-financed projects – as we will see in Chapters 9 to 11.

Contrary to projects under government funding, tier-1 companies in business and industry have obtained practical results. In the engineering and construction domains, for example, designing commercial and residential buildings is greatly facilitated by virtual reality. Through:

- Interactive 3-D graphics, and
- Realtime simulation

civil engineers and architects – as well as the bank(s) financing a real-estate development – can spot problems that may otherwise cause extensive changes and costs during construction. An added advantage is that clients can tour property before it is actually built, making VR a significant help in marketing.

Figure 7.1 presents in a nutshell the analytical methodology underpinning realtime simulation and the visualization associated with a

Figure 7.1 An analytical methodology for realtime simulation and visualization of results

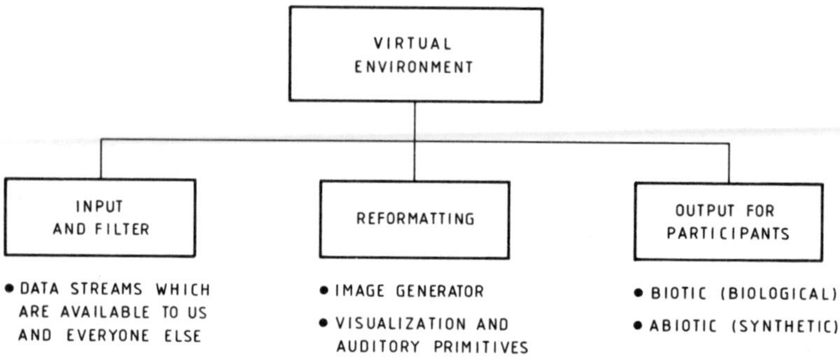

broadband input/output channel. Wideband input and output are two main goals to which new technology addresses itself. They require multimedia reformatting, which today appeals to audio and video streams but tomorrow may cover all of the human senses.

- Navigating between applications and learning how to use graphical user interfaces is tantamount to an integration effort.
- But if software is improperly positioned, installed or utilized, the benefits will be negligible.

The value to be derived from multimedia and virtual reality, as well as realtime simulation or integration, lies in its ability to define and represent a product in its entirety. It is therefore important to access and leverage the information contained in the computer.

In conclusion, the lucrative market which lies ahead has induced network operators, cable companies, software firms and others to position themselves against the forces of the world's largest market: interactive video services. The trend started in America, but it will soon come to Europe as the voice monopoly ends in 1998. When this happens, the video and audio market door will be open for the leading multimedia companies – and this will soon become a global business opportunity.

WHY THE GROUP OF TEN ASSOCIATES FUTURE EMPLOYMENT WITH VIRTUAL REALITY?

Let us first define what is meant by the Group of Ten (G-10). It is an exclusive private club whose only members are the finance ministers or treasury secretaries – as well as the chief executives and central bank governors – of the most industrialized countries in the world: United States, Japan, England, Germany, France, Italy, Canada, Holland, Belgium, Switzerland and Sweden. Luxembourg has an observer status. The Group of Seven (G-7) is a subset in which participate the first seven countries in this list.

Historically, G-7 preceded G-10. The group of Seven started in the mid-1970s with only five members (the first five in the list). Then others were invited to join. What these ladies and gentlemen usually do (apart from toasts and dinners) is a fire-brigade approach to burning financial problems anywhere in the world. But sometimes they also have foresight.

Foresight is what happened mid-February 1995 during the G-10 meeting in Brussels when the world's most industrialized countries decided to sponsor 12 projects on the information superhighway and virtual reality. Another first is that rather than relegating the deliverables at some unspecified time, they asked for prototyping results by mid-May 1995, when the G-10 were to meet again in Toronto. Why the rush?

- The answer is to be found in socio-economic change which has taken place and the impossibility of finding answers through classical methods.
- Many of the conditions that in the past made the lifetime long career possible have disappeared.

Technology evolves so fast that the shelf life of work skills now rarely exceeds 10 years, and even this range is shrinking. It is not only that lower labour costs in other countries have taken away from Western labour markets entire categories of manufacturing jobs; the challenge goes much deeper than job dislocation.

Today, every person in the labour market has to understand this challenge and look at the unexpected twist in his or her career as an opportunity to learn fresh skills. This has to be done before new technology eliminates the need for that person's current expertise –

not after. Fewer and fewer of today's workers should expect to spend their working lives in the same field, let alone the same company.

It is everybody's responsibility to stay up on things and make sure they do not get obsolete. Gone is the time when one lived on the assumption that after finishing school or college there was no challenge in finding full-time work. In fact, he or she could expect to live and work in the same sector of society until retirement.

- This idea of work as an entitlement is practically dead.
- Career paths are no longer as clearly defined as they once were.
- To survive, people have to be cognizant of the rate of change, tolerant of ambiguity and always open to learning new skills.

In fact there is little choice in terms of alternatives, as the spectre of unemployment and even worse of permanent exclusion from the labour force is always present. Lifelong learning is now a basic necessity but the old ways of teaching are too serial, too slow and too inefficient. They do not respond to current needs.

- Because it is a broadband channel and it will be reaching every office, plant and home, the information superhighway may be the answer.
- But it also needs modern tools such as realtime simulation, 3-D visualization and therefore virtual reality.

Every person in Western society has to prepare themselves for the challenge of being self-employed. This will require massive amounts of preparation and training. Otherwise people will lose their ability to hold jobs. They will not merely get disillusioned with corporate life, as has happened with the student revolt of the late 1960s in America during the Vietnam war; they will drop out of society.

It is relatively easy to understand that it is not easy to give up the shelter of the big company, or of big government, and to adapt to the tough rules imposed by the self-employment market. But governments have to explain that there is no alternative to this swing. Traditional methods have resulted in an ever-growing heap of unemployed and of people increasingly excluded from society's goodies.

WAYS AND MEANS FOR AVOIDING THE TRAP OF SELF-DECEPTION

Failure to understand the issues outlined in the previous section is the best way of falling down into a crevasse. But while necessary, simple understanding is not enough. We also have to be proactive in connection to the social and economic turbulence which exists with regard to:

- Knowledge,
- Skills, and
- Employment.

Understanding starts with appreciation of the fact that social frameworks built in the 1930s, the New Deal years, are no longer valid; indeed they crumble. This is equally true of many of those developed during the five decades following World War II. They have collapsed due to their disregard for the dynamism of knowledge, technology and society.

One of the ironies of the 'peace dividend' is that as the end of the cold war has reduced the military threat, social worries in regard to knowledge and employment have been made to stand out.

- Employment is not part of an entitlement, as many people have been led to believe. It never was that way.
- But the chances for employment will be better if we are able to revamp the basis of exchange of knowledge on a global scale.

This is precisely where the interest of the G-10 in regard to the information superhighway and virtual reality, among other advanced projects, comes in. Economic frictions most often are cultural in their roots. They are a matter of knowledge and skills.

Since medieval Europe, needs for better education have been addressed with the help of colleges and universities. In the 15th century, the invention of printing by Johannes Gutenberg (1397–1468 AD) led to the concept of books, newspapers and magazines which we use till today.

The media and methods of education have not changed for over 500 years. Yet, Gutenberg's invention is no longer an adequate process for a mass society of 6.5 billion people – and it has become an abberation when confronted with the very rapid rate of change in employment patterns which has been outlined in the previous section.

- If we wish to find solutions which are adequate for the service society in which we live, then we have to adopt new departures.
- Nobody can assure us today that the information superhighway, multimedia and virtual reality is the solution, but this seems to be the best way to bet.

After all, what are our alternatives? 'Do nothing and let things take care of themselves' is one of them. It has been practiced by some governments all of the time, and by all governments some of the time, but in the employment front it gave absolutely no results.

'Do only a little here or there', also known as the policy of patching, has not reversed the deteriorating employment situation either – particularly among the young. Semi-measures and half-baked approaches never work, if for no other reason than because they lack imagination.

During the post-World War II years, the socialist-leaning governments of Western Europe were supposed to have paid great attention to social problems. Low wages were pushed up through readjustments of the minimum wage and sometimes by big jumps.

At the same time, unemployment insurance (a misnomer) professed that unemployed people got quite a good deal by getting paid whether they worked or not. This distorted the social and economic value with the result that:

- With globalization, Western societies priced their labour out of the market, compared to the high skills but low wages of Eastern Europe and Asia.
- Not only is Western European unemployment at record level, but also if a person never had a job he stands a good chance never to get one.

The French have developed a now vocabulary to deal with the situation: *exclusion*. It is a word in vogue. In other countries, like Denmark, social problems have been tackled by creating marginal work which satisfies nobody.

This is indeed an example of half-baked solutions. The Danish government gets the unemployed to look after the old and the sick.

Since about one in seven of the population is out of work and paid enormous sums to stay out of the job stream, this scheme adds up to a very expensive way of providing social assistance.

France calculates that each person unemployed weighs on the national budget an awesome Fr. 126 000 ($25 000) per year. That's red ink. Unemployment insurance costs big money and this means higher taxes for everybody else. Those who work have to pay for every retired and every unemployed person, and above that for the:

- Health insurance, and
- Inefficient administration of social services.

This is the so-called *social net* whose weight is crashing everybody supporting it. Such silly business happens because no politician has the guts to say that chronic welfare beneficiaries are a bunch of layabouts and that it is not worth increasing taxes to keep on paying them silly money.

Not only do the current semi-measures not take the unemployed off the hook, but they also swallow their ranks. Hence new departures should be welcome, particularly if they are of an imaginative nature.

- But are the G-10 government serious about starting on a totally new road?
- In terms of deliverables, judging from what has happened with initiatives such as HPCC in the United States, ESPRIT in Europe and ICOT in Japan, the answer is at best ambiguous and at worst negative.

We will be taking a very close and critical look at HPCC and other projects from Chapter 9 to Chapter 11, where the reader will find plenty of evidence on why and how government-sponsored projects are mismanaged. First, however, it is appropriate to give the reader a good practical background on what virtual reality is and what it can do.

THE EXPANDING HORIZON OF 3-D REALTIME COMPUTER GRAPHICS

The best way of looking at *virtual reality* is in a practical, implementation-oriented sense. For instance, as an extension of 3-dimensional realtime computer graphics, whether with or without immersion. To its proponents, VR promises to be a revolutionary experience in information handling – a statement which has to be examined carefully if its affirmation (or negation) is to be convincing.

Definitions of what virtual reality is and what it is not vary, but on the bottom line what these definitions say is that in essence what we call VR is a way of enabling information users to:

- Participate directly in a realtime simulation, and
- Feel as though part of 3-dimensional environments generated by computers.

In this sense, as we have already seen virtual reality solves the human–computer interface problem by providing a broadband input/output platform. That is why, in all likelihood, over the next few years it will have major impact in key areas of professional and daily life.

Not everything is of course totally new. As we will see in the next section when we talk of VR applications in engineering, the predecessor has been 3-D colour graphics. The now evolving principles of practical VR applications help in creating a new generation of interfaces for users of large, multimedia databases. There are three approaches in reaching this goal:

1. A participative solution (see also an example discussed later in this chapter),
2. The immersive VR environment, and
3. A 'through the window' approach.

The more advanced may be considered to be the *participative*, or active-person solution, in which one sees his own image on the computer screen, interacting with the virtual world. The sophistication which can be provided is significant, limited only by the software needed to generate it and the processing power available on high-performance computers to bring it to life.

Figure 7.2 Realtime ski simulation through virtual reality*

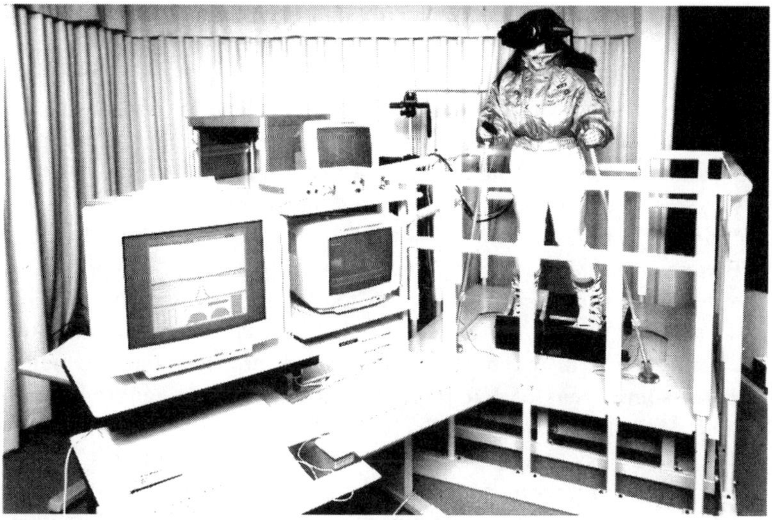

* The mathematical model underpining the VR implementation has been developed by NEC.

Short of fully participative characteristics, a similarly advanced virtual-reality solution is the immersive approach. With this one enters the virtual world wearing data garments such as a head-mounted display and data glove, to map his senses directly into the digital environment and vice versa.

- Head mounted displays provide a stereoscopic image of the virtual world,
- A built-in audio system assures the accompanying audio sensation, and
- The data gloves help in the 'put it there' way of pointing at the screen.

Taken from a simulated realtime environment developed by Nippon Electric, Figure 7.2 shows a skier with head-mounted display. The subject has all the sensation of going downhill on a slope but without the risk of breaking a leg. This is a good example of how virtual reality can be used for training purposes.

For its part, the data glove projects an image of the user's hand. By manipulating the glove, the user can directly interact with the computer-generated reality, handling and moving objects or using gestures as commands. By means of this set-up one can move, even fly, through the virtual world.

The drawbacks of this approach are the head-mounted display and dataglove. The former has evolved from astronaut training in the 1960s, but it is cumbersome and of limited use in a business environment: the latter from research done at MIT in the early 1970s for the US Navy.

Business people with experience in virtual reality applications complain that the classical model of the VR helmet is clumsy, heavy and uncomfortable – but as we will see in a later section, when we discuss resolution and field of view, there are new developments. Still, in some areas of VR implementation, such as banking, head-mounted displays cannot really be used.

The third, less-complex approach of the three – but still a great improvement over other methods – is the so-called 'through the window' solution. With it one experiences virtual reality by looking at the three-dimensional world through a computer screen. The user steers through the environment using a spaceball, flying mouse or some other control device with sufficient degrees of freedom (explained in Chorafas and Steinmann, 1995).

Directly connected to these three VR presentations is the world of the new terminals for multimedia handling. These new terminals are evolving mainly in two directions. The one is towards small-sized, lightweight, location-free devices, aimed at allowing users to talk with:

- Anybody,
- Anywhere,
- At any time.

The other direction is towards the integration of 3-D, or nearly 3-D, multimedia features, bringing moving image into conversations between persons through a broadband communication network. This approach also aims to present a large amount of processed and edited information in a form most easily understandable by users.

Readers knowledgeable in VR would say, not without reason, that we are really talking of two things at the same time: *virtual-reality* terminals (which we have already briefly examined) and *personal-communications* terminals, more appropriate to the location-free features of the information superhighway.

Today, such a statement would be valid. But tomorrow, it would be wrong because of the convergence of technologies. An ongoing transition focuses on location-free terminals and takes place as mobile communications are evolving from analogue to digital format. The goal is that of:

- Obtaining multimedia solutions with the ability to transmit and receive voice, data and image information.
- Eventually doing so by exploiting the broadband input/output channel that virtual reality offers.

In principle, the ultimate goal in communications is terminals which allow immediate links with anybody, anywhere and at any time – including a virtual setting. For instance, for teleconferencing purposes.

As the telecommunications network evolves from analogue and pseudo-digital ISDN type – both of which rest on plain old telephone service (POTS) – towards broadband, portable memory devices will be added to the terminal. These will generate, store, edit and display all kinds of media, ranging from voice, data, image and standard video to HDTV and VR.

The new generation of terminals will also be able to send and receive a great variety of information over a broadband network. Clearly enough, there are many problems that remain to be solved in communications using multiple media, including how a session is to be formed. Nobody said that all necessary devices and solutions are on hand – but they are evolving and the 12 projects promoted by the Group of Ten aim to accelerate this evolution.

FROM REALTIME SIMULATION TO VIRTUAL-REALITY APPLICATIONS

'A virtual environment', said MIT's Dr David Zeltzer, 'is a better platform for computer applications in science, engineering and business.' What is new with virtual-reality implementation is that the workstation of the end-user has progressed in a way that we no longer need millions of dollars to make realistic and interesting the results of:

- Realtime simulation, and
- 3-D computer graphics.

Technology is now providing miniature stereo displays at an affordable cost. As the Santa Fe Institute was to underline in the course of our meeting, the whole issue of implementing virtual reality is to significantly improve upon presentation of information – not its processing.

Whether in a virtual or in a real environment, a realistic presentation of information can take many aspects all of them having to do with different degrees of sophistication in visualization. Let me, however, add a warning prior to looking more carefully into technical issues. The solution we are after must be flexible, and we should never box ourselves into something parochial and limited.

In 1966 I was participating in a seminar on what was then high technology, organized by the Chase Manhattan bank and chaired by Dr David Rockefeller. 'Do you know why, in a financial sense, Concorde will not fly?' asked Dr Rockefeller. 'Because it is loaded with unnecessary costs', I answered – a response which was given to me a week earlier in a meeting near Bristol with British Aerospace, Concorde's co-designer.

'This, too', said David Rockefeller, 'but the Number 1 reason is that its design is boxed-in. It comes only in one size. It cannot become smaller and it cannot become bigger. Hence it cannot amortize big money spent on R&D.' Figure 7.3 dramatizes this reference. A first-class example of a *flexible design* is Boeing's 707 which was downsized to 723 and 737; and grew to become the 747 aircraft.

Designers will be well-advised to pay respect to this concept. Specifications should see to it that a product is able to be downsized and to grow. This principle is particularly important in an expanding field like virtual reality, which owes much of its background to 3-D graphics and simulation – hence in processes which are inherently flexible.

During the cold-war years, realtime simulation has been extensively used by the military to test novel situations, study unknowns in the battlefield, and for training reasons. Today civilian-oriented work puts emphasis on *applications building* which poses a number of challenges in:

Figure 7.3 The design of complex engineering products should be flexible, not locked into a box

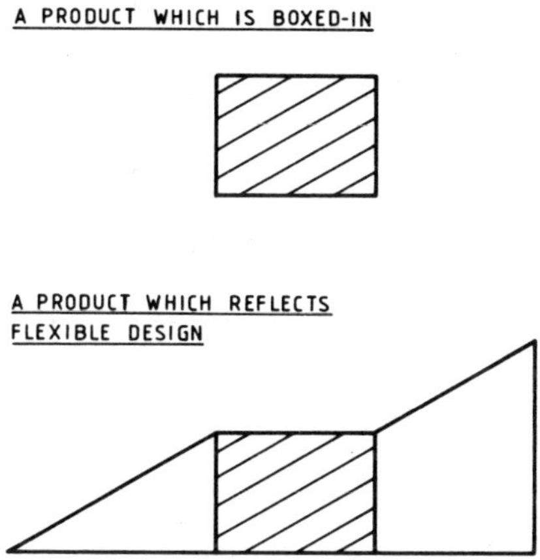

- Systems engineering,
- Human interfaces, and
- Applications development.

The more advanced is the solution, the better able we are to cut inside the problem, emulating behaviour through technology. At the same time, the more we face in an able manner realtime simulation issues, the more the deliverables permit us to extend the human senses.

Clearly enough, the factors of success do not only lie in building simulators, by using algorithms, heuristics, knowledge engineering and the implementation of 3-D graphics. A most significant part of the job is in collecting and filtering data to be used in providing effective visualization solutions. The challenge comes in stages.

- Designing an aircraft carrier or a financial system is a much larger problem than a 'one product at a time' orientation.
- Complex systems affect the *whole process* and the *solution space* must take this into full account.

Figure 7.4 Virtual-reality projects can benefit from a layered approach to their development and use

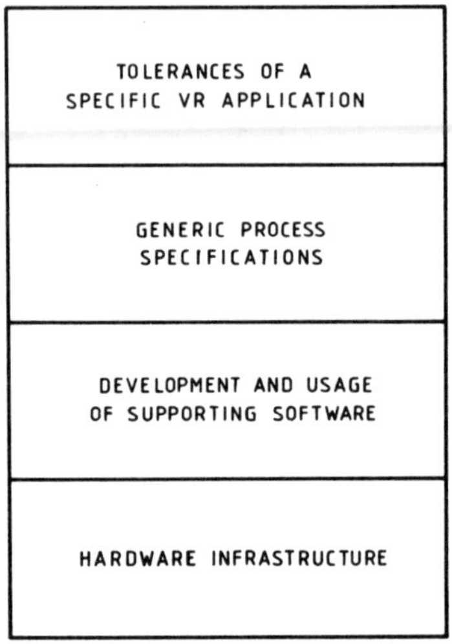

The best way to approach virtual-reality projects is in a layered fashion by specific levels of reference, as shown in Figure 7.4. The first step is to interactively establish, with the end-user, the tolerances characterizing a specific application. This is a concept well-known from engineering design.

Application-specific functions rest on the facilities provided by the next layer: the process specifications which are generic, and can support one or more VR projects. Below this level comes software development including off-the-shelf packages. At the bottom layer are the hardware resources. The logical infrastructure of layer No. 3 includes:

- Higher-level graphics tools with a set of libraries and primitive commands.
- An object-oriented operating system and database-management system (see also Chorafas and Steinman, 1993a).
- Input/ouput supports, utilities and programming languages.

As already noted, beyond the better-known simulation and graphics aspects lies the problem of realtime data analysis. Cornerstone to this effort is the creation of an intelligent database structure which will be *rendered* and presented to the screen. *Rendering* can be visual, audio or tactile.

Among the technical problems confronting the application is feeding into the database generator. This may pose bandwidth problems, as with *generic visualization* some applications create a double-digit number of files per second.

One of the challenges with generic visualization is the generation of networked computer-supported collaborative work. All actions in one screen must be reflected on the others, but at the same time everybody may need to have his own 3-D view. Let us take an example from engineering.

CAPITALIZING ON THE VIRTUAL ENVIRONMENT FOR ENGINEERING DESIGN, FIELD OPERATIONS AND TRAINING

Say that a design engineer crosses the gateway into the world of virtual-product design. Using a realtime computer-generated environment, he reaches out and grabs a 24 000-volt powerline, pulls it out of the network in terms of power transmission, then studies the dynamics of the resulting sub-network.

Far fetched? Not at all. It is an example of the *participative* solution we have examined earlier. This solution has been effectively applied by the University of Tokyo in a project sponsored by Tokyo Electric Power, Japan's foremost power production and distribution company.

- Such applications are systems breakthroughs.
- They are effectively capitalizing on major advances.
- Among the elements underpinning them are simulation, resolution and field of view (FOV).

What does *resolution* mean? The human senses can view about 25 000 pixels horizontally and almost that many vertically. Current head-mounted displays which incorporate liquid crystals offer a resolution of approximately 800 x 600. This discrepancy accounts for the somehow primitive appearance of objects, which is however in the process of being corrected.

In connection to *field of view* humans can approximately process a 170 degrees wide by 130 degrees high FOV. By contrast, head displays commonly in the market offer FOV of 110 by 50. One of the problems in matching the human senses is that displays with a wide FOV must often sacrifice high resolution.

- In a resolution sense, the wider the FOV, the more pixels are needed to display that area.
- At the same time, advanced applications projects require a wide field of view to give the participant a sense of immersion.

Some applications, however, circumvent the limitation resulting from a combination of wide FOV and enough resolution to match the resolving capabilities of our eyes, providing other advantages inherent in realtime simulation like the participative capability of the Tokyo Electric Power implementation.

Another example of virtual-reality implementation in power production and distribution comes from Italy, where ENEL, the state-controlled power company has ongoing projects to use VR in connection to training as well as in production planning. Projects include:

- Training engineers, particularly for network parts which are difficult to access.
- Interactively developing skills for security purposes, including the presentation of options for operations control.
- Doing other multimedia applications like walkthroughs on a power-production factory floor and associated assignments.

In terms of VR experience, a great deal is being gained in a cross-country and cross-industry sense. Some of the applications, such as the manipulation of difficult or dangerous engines in a power-production plant have been already accomplished in other industries – for example in connection to earth-moving equipment.

A real-life implementation example is given in Figure 7.5. It comes from Japan and concerns a virtual-reality application done by Fujita, one of the better-known construction companies:

Figure 7.5 A practical implementation of virtual reality in the construction business

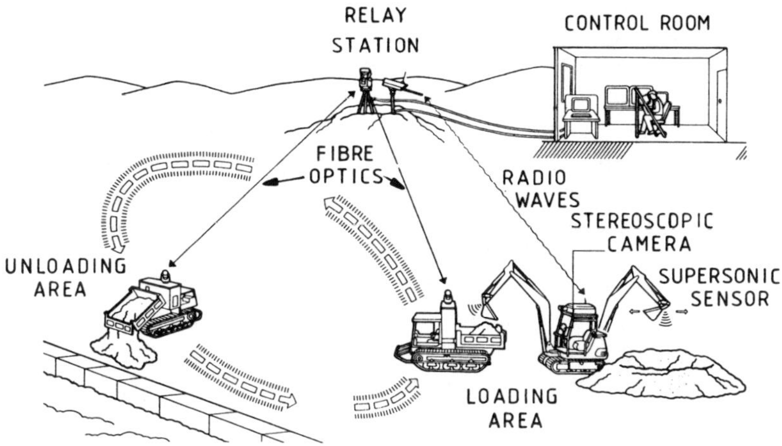

- Through VR input/output, the operator in the control room manipulates heavy machinery at the loading site.
- A whole infrastructure is, however, necessary to support this application, involving a relay station, fibre optics, stereoscopic cameras and supersonic sensors.

Other VR applications are made for larger audiences, particularly for training and demonstration reasons. Several projects have been part of the broader perspective offered by research, development and *implementation* (RD&I), rather than R&D alone.

For instance, the Cave Automatic Virtual Environment (CAVE) is a VR theatre developed by the Electronic Visualization Laboratory at the University of Illinois, Chicago. It is designed to give the impression of a 3-D interaction space so that users feel they are *inside the database*.

Carterpillar is one of the companies which put the CAVE facilities to profitable use. The application employs a 10-cubic-feet room with projection displays at the walls allowing product designers to operate a virtual machine. They are:

- Testing alternative machine designs,
- While driving through a virtual proving ground.

Hydraulic actuator systems, for example, are simulated in realtime permitting operational effectiveness assessment. This makes it feasible for designers to quickly evaluate a number of options and get a feel for their artefacts early in the design cycle. In essence, this VR implementation is a sophisticated approach to *database mining*.

A CAVE application involves a designated *driver* who uses a control device identifying a particular area of the database to be explored. There is also a collection of passengers who are able to observe the simulation which includes:

- Adverse weather conditions such as snow storms,
- Different gravitational waves, and
- Topological surface deformation.

What is learned in terms of benefits provided by VR in electrical and mechanical engineering, can cross-fertilize other fields. Interactive computational finance is another important applications domain, and therefore exploring the benefits to be derived by realtime simulation, with associated 3-D colour visualization, makes sense.

COMPUTER-AIDED DESIGN AND VIRTUAL REALITY IN ENGINEERING AND MANUFACTURING

Much like computer-aided design (CAD), developed in the 1960s in the laboratory of Dr Ivan Sutherland at MIT, virtual reality has come out of academia and the research laboratories into industry. We have seen a number of examples in the previous section – but it is also interesting to take notice that there exist several parallels between the two processes, CAD and VR.

One of the similitudes leading to analogical reasoning is the original scarcity of tools. When CAD was first envisioned, computer graphics were not widely available and the enthusiasm for CAD existed primarily in universities. Yet large manufacturers, like General Motors, saw the potential and capitalized on the concept.

- To apply CAD notions to the design of automobiles and trucks,

GM had to start from scratch, developing by itself the entire CAD system.
- But as the value of this process was proved, other major manufacturers used CAD, leading to the development of commercial hardware and software.

Quite similarly today, major automotive, aerospace, consumer products and other manufacturers have recognized the value of virtual-reality interfaces in terms of improving their products and processes. Particularly processes such as:

- Styling,
- Engineering, and
- Manufacturing.

Therefore, they have set projects in place to try and harness the value of VR. Like CAD in the 1960s, there are many difficult lessons that have to be learned before VR can become a popular element of the standard product development and manufacturing cycles.

The first basic lesson concerns the need for new concepts in the use of VR. The second revolves around the requirement for leveraging the benefits which can be obtained from interactive computer graphics – which is a process already in wider use as Chapter 1 has explained.

In a number of ongoing VR projects, systems refinement allows particular areas of the displayed data to be explored in greater detail than it would otherwise have been possible. One of the key lessons learned from developing non-traditional applications which operate online is that virtual reality need not be done at the multi-million dollar level to realize useful results.

Just as important is to take notice that many virtual-reality prototypes are developed in a short timeframe. Also, in the majority of cases where prototypes in a VR environment proved to be instrumental to newly-designed engineering products, the cost has been much less than a comparable physical prototype.

As a result of a full 3-D interactivity the user perceives the simulated world directly. This interactively available perspective is clearly superior to any previous alternative because:

- It permits the user to rely on innate spatial-perception abilities, and

- Makes it easier to comprehend what we see and plan what we do.

Alert engineering designers and financial analysts get a better feel of the objects they deal with when they reach out to handle them. They can better perceive size, distance, frequency and other crucial variables.

Once companies recognize the operational benefits of using such systems, and the deliverables live up to these expectations – virtual reality will become a major competitive weapon. It will also change in a significant way the manner in which we:

- Learn,
- Think, and
- Work.

In essence, the projects connected to the information superhighway and VR, financed by the Group of Ten, have the dual scope of popularizing these new processes and of providing a stimulus for the development of more efficient tools. Under this dual aspect should be seen their contribution to opening up new employment opportunities.

HOW NEW DEVELOPMENTS ARE BREAKING OLD BOTTLENECKS

As graphics hardware and software is developed and incorporated into the everyday computing tools used by designers, engineers, financial analysts and other professionals, new applications require solutions which employ greater:

- Interactivity,
- Field of view, and
- Rendering quality.

In turn, this calls for still better VR tools than those currently available, as well as for intelligent networks and database-mining capabilities. These requirements are related among themselves, as we have seen earlier.

There is also the need to break the bottleneck presented by classical head-mounted displays. One promising new technology is a laser-emitting device attached to a pair of eyeglasses that transmits images directly onto the wearer's retina.

In terms of size and weight, makers of head-mounted displays are also making progress. As a result, the industry has seen a proliferation of lightweight choices, some weighing as little as 3 ounces.

Besides this, head mounts are not the only possible solution. Many experts believe that displays of the future will not be head-mounted but wide flat panels with 3-D effects able to support realtime scenes required to make the presentation realistic to the human eye.

The area of visualization is advancing rapidly. Chips that optimize the display of 3-D images are under test and will soon be used for graphics and video boards for personal computers. Also in terms of horsepower it is now possible to get sufficient performance out of computers.

- VR performance is measured in frames per second and polygons per second, both factors constrained by the power of the computer running the simulator.
- A frame rate of 30 frames per second is generally recommended to convey the feeling of smooth motion, and is currently supported by high-end systems.

Another important factor to be kept in perspective, because it influences unit costs, is that the number of implementation domains is expanding. Financial analysts can use high-performance computers to generate tick-by-tick trading patterns – just like aerospace engineers create computational fluid-dynamics patterns.

The analysis of *high-frequency financial data* has been a bottleneck since the beginning of analytical computational finance, in the mid-1980s. As new tools for *qualitative* analysis (see also Chorafas, 1994) and realtime visualization come onstream, this bottleneck is broken.

Rocket scientists are now applying to finance the experience they have gained in nuclear engineering and in realtime simulation of the aero- or hydro-dynamic qualities of a new vehicle or hull design. They also address the issues connected to necessary technology transfer from engineering to finance.

Both in the financial analysis and in the aerospace industry, virtual reality offers a first-person perspective. As we saw in the preceding sections, it immerses the user into the computer generated space through 3-D input/output devices.

- These permit the user to interact with objects in the virtual environment.
- Therefore, they shatter the barrier formed by making people interact with 3-D objects on a 2-D screen.

The employment aftermaths are not difficult to comprehend. An industry is built when there are a lot of people using it as mainstream in their day-to-day operations, and for that to come about performance and price are key factors – but also the ability to convince users that the new business opportunities are here to stay. The most important ingredient of business is confidence.

INCREASING THE BENEFITS TO BE GAINED FROM CONCURRENT ENGINEERING OPERATIONS

Experience with virtual reality demonstrates how much easier it is to understand spatial relationships between computer-rendered 3-D objects. By immersing themselves into the space generated through realtime simulation, users can employ their professional experience to better interpret and understand spatial relationships in visualization.

Therefore, it comes as no surprise that today tier-1 companies leverage computer graphics for just about every stage of design, manufacturing and field operations. Immersive interfaces can be applied to every one of these steps, but also non-immersive VR solutions provide a significant advantage.

Practically everyone with experience in VR applications seems to agree that in the not too distant future virtual reality will profoundly change the way products are conceived, designed, tested, manufactured and maintained. Evidence to this is the fact that most major automotive and aerospace companies have taken the VR ball and run with it not only in their laboratories but also in the production environment.

The exploitation of the benefits virtual reality can offer in the domain of *concurrent engineering* has become a focal point, because networked collaborative work is most crucial in a number of projects. For instance,

- Shortening the time to production and delivery (time to market),
- Cutting the fat from a product or process by reducing cost in a company-wide sense,
- Minimizing possible damage due to risk and exposure, which is often not unattended to.

Projects along this line of reference benefit from database mining and use VR to make more efficient the visualization of obtained results. The ability to promote cooperative work is a major reason why engineering companies, simulation laboratories, plant design companies and financial institutions are moving into VR.

In engineering, for instance, collaborative work for product visualization permits us to look at the same product from different viewpoints, communicating in realtime on details. In finance, networked collaborative work may regard portfolio evaluation in different market settings, as well as hedging strategies.

The role a virtual-reality approach can play in both examples is that of a broadband input/output implemented at an analytical level, which can present critical information on processes, databases, interaction of objects and alarm conditions. As it has already been underlined, this amounts to a significant modernization of I/O interfaces.

- Electrical and mechanical design and virtual reality can be a perfect match.
- The change from the projection paradigm into an interactive VR environment can be profound, with important cultural aftermaths.

When engineering designers can see and use VR tools, getting the feedback needed to improve their concepts, there will be significant changes in the way products are designed. Companies will be able to turn products around faster and make them more reliable, because their experts can understand and interact with them better.

In conclusion, critical to the whole process of innovation we have examined in this chapter is the ability to generate a virtual environment. This is essentially what the projects sponsored by the Group of Ten try to support, but user organizations also have considerable work to contribute. The online databases, algorithms, heuristics, knowledge-engineering artefacts, other analytical tools and simulators are examples of such contributions.

Today the major industrial companies are investing in developing their own mock-ups, because commodity CAD systems are not yet doing the whole lot of needed services. Therefore, they are learning lessons that will translate into a first-hand experience as the broadband channels of the information superhighway come into existence.

The solutions to be adopted can lead to a great demand for channel capacity, therefore VR and the information superhighway can be supportive of one another. They are both on integral part of the restructuring revolution necessary to graduate to the post-industrial society.

8 The GreenSpace Project and the Future of Multimedia

INTRODUCTION

While multimedia is a term which, over the last ten years, has grown from a concept to a number of successful applications, its current use is already judged as too limited in terms of the projected needs of business and industry. This is true all the way from interfaces, to servers, networks and protocols – which are the building blocks to the practical application of virtual reality as discussed in Chapter 7 (see also Chorafas, 1994a, and Chorafas and Steinmann, 1995).

Along this frame of reference, the purpose of the GreenSpace project undertaken by the Fujitsu Research Institute and the Human Interface Technology Laboratory of the University of Washington, has been to develop and demonstrate a new and practical communications solution. Multiple participants are immersed and feel a sense of presence within a common interactive environment of visual, aural and tactile information shared among them.

Solutions to the challenging problems posed by an enlarged implementation perspective, require that system designers return to the fundamentals all the way to the background where the action starts. In GreenSpace project jargon, this basic background is known as *virtual common* and is intended to provide a transportation system for natural senses and motor behaviour of multiple users. The goal is to:

- Immerse simultaneously up to 100 people in the same virtual common, and
- Do so whether the participants are nearby or separated by great distances.

The GreenSpace project plans to grow a *virtual network* fabric starting with a local network and expanding to a regional setting.

The next step is covering the Pacific rim by connecting Seattle to Tokyo. Many companies participate who so far have been competitors but now have common goals to meet.

The solution which is underway features an architectural framework based on detailed examination of the concepts underlying the individual applications. Further-out are projects which will use the currently virtual common and virtual network. This calls for an examination of the implementation structure, including new communications approaches; also for careful definition of the technology participants in GreenSpace now use and will use in the future.

Once present and future requirements have been defined, the virtual common and virtual network will be designed so as to abide by systems principles, including a significant level of complexity. What a system delivers can be expressed in terms of service objectives which evolve over time, but in their current form are described in the next two sections.

GOALS AND TIMETABLE OF THE GREENSPACE PROJECT

A basic principle in the development of a modern system architecture is that of focusing not only on how the network is structured but also how it operates under a number of different scenarios. These must be analyzed through functions and objectives. Another critical reference is how the network will stand in maintenance terms, including the problems characterizing cross-border implementation.

Figure 8.1 brings these issues under perspective by providing a unified vision of design, management and operations as indivisible phases of any network project. Each presents both development and application issues which, in GreenSpace, have been mapped into five consecutive phases. Phase 1 started in January 1993, lasted about two months and focused on concept development. This practically laid the foundations for what followed therafter. Phase 2 was divided into two parts,

- The one addressed the initial design study,
- The other the so-called expanded design study.

Phase 2 ran from April 1993 to March 1994, and it has been succeeded by Phase 3 which began in April 1994 and ended in early

Figure 8.1 Design management and operations are indivisible phases of any network project

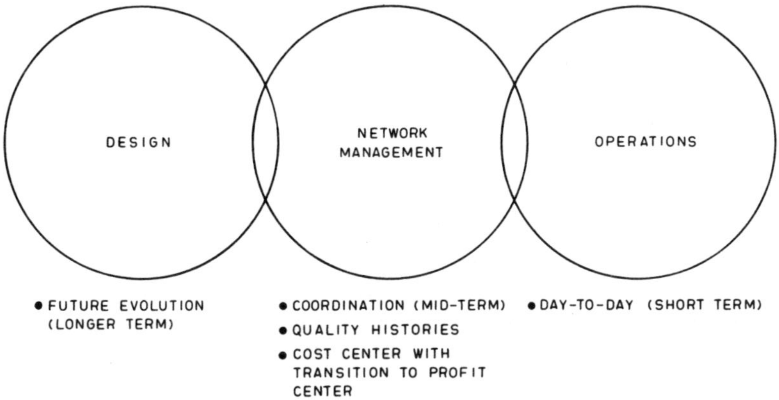

1995. Its goal has been what became known as the Initial Testbed Development. Beyond this came Phase 4, targetting an expanded testbed development and lasting till early 1996.

The goal of Phase 5, which will start at the completion of Phase 4, is that of developing and implementing an International Testbed. Another objective is the establishment of a GreenSpace Centre – a project which will probably continue till mid-1997, give or take a few months.

On the American side, funding is provided by the National Science Foundation, Advanced Research Projects Agency (ARPA), NASA and the National Oceanographics and Atmospheric Administration. Beyond these government agencies participate Bellcore, US West, AT&T Network Systems, MCI, Sprint, Silicon Graphics, Evans and Sutherland, Sun Microsystems, DEC, Intel, Virtual Vision, Virtual Reality, Kendall Square Research (KSR), Ford Motor Company and Boeing.

There are also a number of participating universities, like the University of Washington, MIT, North Carolina, Rutgers, Virginia, Illinois and Central Florida. The University of Tokyo and Tokyo Metropolitan Institute of Technology are players on the Japanese side. Other Japanese participants are Toyota and Sharp – and of course Fujitsu which is the project co-manager.

There must be a reason why so many sponsors from government, industry and academia put money and trust in GreenSpace. The principal reason is fulfillment of the stated object of a far-out communications solution characterized by:

- A virtual common, and
- Virtual network.

A second purpose of the GreenSpace project is to develop the concept of an international centre for exploring the future of the information society. In fact, the title *GreenSpace* was chosen to communicate the notion of an open, common space where new ideas and concepts can grow through participants:

- Brought together via sensory immersion, and
- Sharing a networked virtual information environment.

Virtual interfaces and the environments they generate help in providing new perspectives for communicating situation awareness. They make it feasible to build a landscape, which can be seen as a centre without walls, able to represent the interests and involvement of world citizens without their leaving home.

The projected activities of this fully-distributed landscape include education, scholarly research and development, as well as concept demonstration using information media. *Commons* have been defined which permit the exploration of many applications domains, among them:

- A virtual world for teaching purposes, and
- A virtual corporation setting (see also Chapter 7).

So far, through its first four phases, the GreenSpace programme has helped to develop and demonstrate an advanced interactive concept empowered by a broadband telecommunications network. This effort has effectively combined emerging virtual interface technologies with fibre optics to serve a *virtual community* defined as:

- Places experienced with the senses, but

- Created electronically using virtual interface components linking the participants.

GreenSpace generates a three-dimensional virtual world of sight and sound within which participants can interact with each other and communicate with complex distributed databases (see also the discussion later in this chapter of multimedia databases). A unique aspect of this landscape is that it is *virtual*.

The project addresses many of the major barriers in cross-border communication, such as the differences in language. Every nation is torn between trying to preserve the national heritage versus the alternative of learning a 'universal' language which permits an effective interaction among people.

Therefore, one of the crucial aspects of the project is that of natural languages. New ones are being studied, based upon spatial experiences which do not translate directly into established language worlds. An example is a new language based upon a spatial approach and icons.

A number of relevant technologies have been experimented with, and are in the process of being productized. They include *agents*, video broadcasts over fibre-optic cable, video conferencing, distributed programming, video-on-demand, text-based and graphics-based multi-participant gaming, as well as video shopping. Other technologies are interactive messaging, distributed computing platforms, and realtime simulations in a virtual environment.

GreenSpace considers video conferencing experiences as being very important to society beyond the year 2000. This explains why many telecommunications companies are teaming up with cable network and computer vendors to produce true, interactive teleconferencing systems. The project is:

- Using a window that all callers can see and simultaneously draw on, type and paste images into, and is
- Allowing participants to see and hear each other as well as to draw on a common blackboard.

Distributed programming techniques are a hot topic of research, because they constitute the best solution in the distributed landscape of the virtual network. GreenSpace developers see the potential for increased productivity in highly-interactive, distributed

Figure 8.2 Both greater bandwidth and higher levels of intelligence will characterise future networks

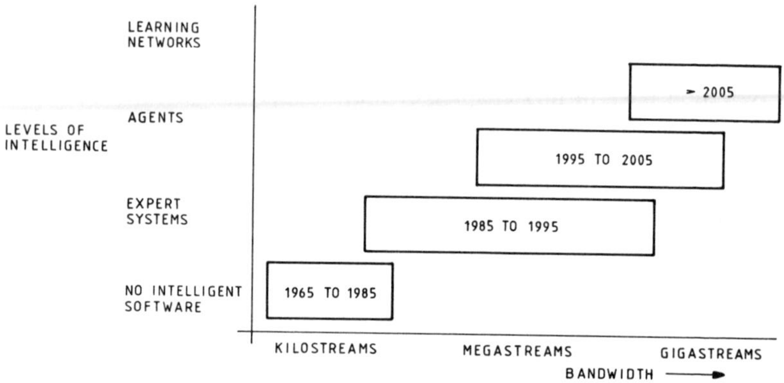

computing environments and they are creating the tools necessary for building large-scale simulations.

NETWORKS WITH GREATER BANDWIDTH AND HIGHER LEVELS OF INTELLIGENCE

Given the amount of data that a person or a company can be exposed to in today's networks, and especially those of the years to come, we need intelligent assistants to help us navigate through the information elements and filter out digital data streams. This work is done by *agents*, who know what their master likes or needs to have for specific purposes.

While most emphasis is being placed today on bandwidth, as Figure 8.2 shows there is in reality a dual evolution involving both bandwidth and the level of intelligence embedded into networks – a fact of which we have spoken on several occasions. This figure dramatizes four epochs:

- Telephony as we presently know it, with kilostreams, which characterized networks prior to 1965.

- Current multimedia solutions resting on digital links, using megastreams and having integrated expert systems at the nodes.
- Multimedia networks of the next ten years which will use gigastreams and a growing population of agents with significant intelligence.
- Network designs ten years down the line with very sophisticated artefacts able to learn as they work; and parallel to this, channel capacity with three digit gigabits per second.

The evolution characterizing the last two points is based on research findings among vendors and in the user community, suggesting that end-users will increasingly require system capabilities based on the growing requirements of the task they wish to perform.

Technology must enable the participants to effectively employ their natural skills while working on the network, and should as well provide a performance level which will be steadily growing. While the solutions we are after will be increasingly based on fuzzy engineering, genetic algorithms and heuristics, what the end-user perceives must be unambiguous, easy to learn, and provide a high bandwidth bridge to the brain – without consuming excessive resources.

The system must dynamically adapt to the given task, be reliable, be tailored to individual approaches, operate with a natural semantic language, and carry high semantic content. As far as I can see:

- Such solutions will typically aim at a spatial representation of data and points of view.
- They will afford both macroscopic and microscopic approaches, as well as lead to dynamic information filtering.

Both GreenSpace and other similar projects in the making will increasingly support a sense of presence in realspace with intuitive navigation. *Realspace* is the realtime exploitation of a distributed space environment into which, through simulation models and broadband telecommunications, we try to compress events taking place globally to one *virtual point.*

- This virtual point should be comfortable and familiar to the user – yet interesting.
- The construction of any virtual environment must be functional and enabling.

There should be a consistent use of metaphors wherever they are used, and an object-oriented approach will dominate the design of the virtual environment. Developers and participants will necessarily work very closely together, operating iteratively on each implementation scenario and its extensions.

To make possible realspace simulation, the transmission of information between networked workstations must have minimum latency and inter-arrival time. There must also be adequate continuous bandwidth and transient bandwidth on demand, so that a guaranteed quality of service can be defined, allocated and maintained.

Whether steady or bursty, transfer of digital datastreams must be reliable. Lost or mangled packets must be recovered either through redundancy in the encoding of the transfer, or explicit re-send requests from the receiver – actuated through agents.

To turn this scenario into reality, knowledge-enriched tools must be available at many levels. Aside from the infrastructure equipment such as new types of software, computing platforms, switching hubs, maintenance and test gear, a number of high-level system tools are required. For instance:

- A suite of authoring tools for environment builders to create useful realspace simulations.
- Individualized means enabling each participant to accomplish his task, given his expertise.
- Benchmarking suites necessary to handle pledges and performance requirements associated with different scenarios.

These requirements will evolve, and the techniques developed will be manifested in them to a greater degree than today's notions or demonstrations seem to suggest. The network itself must provide adequate services for scalable sessions, which is a most challenging design issue.

CHARACTERISTICS OF AN INCREASINGLY MORE SOPHISTICATED NETWORK

What the previous section has outlined is doable, as the evolution we have experienced during the last dozen years shows in terms of practical applications of artificial intelligence documents (see Chorafas,

Figure 8.3 Agents are networked, mature knowledge artefacts, whose evolution can be described in 3-D

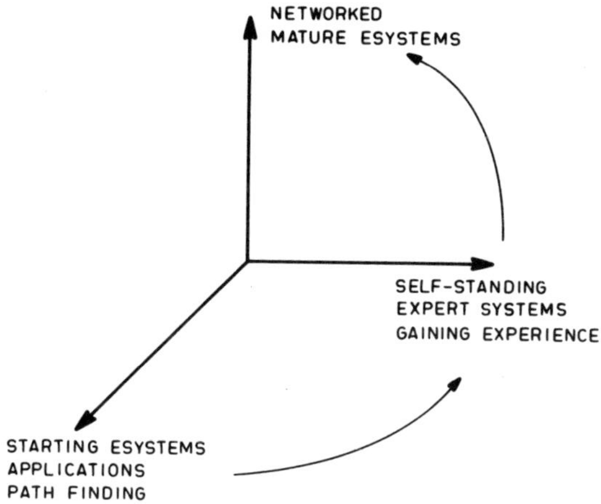

1992). As an example, Figure 8.3 presents in a 3-dimensional graph the significant change which has taken place in connection to expert systems (E-systems) – from the first rudimentary artefacts to the current use of agents.

It goes without saying that system performance needs to be continually assessed as it is developed, installed, tested and upgraded. System metrics include transmission and switching speed as well as reliability, data throughput rates, display-update lag time, and so on.

Critical is the identification of system bottlenecks and other weaknesses. A growing number of specialized agents will be needed as network, server and workstation performance will need to be examined in:

- Qualitative, and
- Quantitative terms.

Not only network intelligence in a user-oriented sense, but also component behaviour and system reliability need to be assessed on an *ad hoc* and an ongoing basis. This means that performance and reliability have to be monitored and logged by the field engineers at each site, along with any unanticipated problems and their apparent or suggested solutions. Hence, the need for:

- Online quality databases, dedicated to network management, and
- Agents capable of mining these databases, providing exception reporting through visualization.

Quite similarly, since ease of use is a major determinant of system acceptability, virtual-reality software must not only be available but also easy to employ and comprehensive – even intuitive for all users.

Hence a particular effort should be made to assure usability, with questions of professional interests, research methods, and required measurement instruments determined during the design study stage of the project – but steadily updated thereafter. This is, among other key issues, one of the goals of GreenSpace.

This principle is simple: it makes absolutely no sense to perform such jobs in a manual manner, even if this was feasible under a very heavy information load. Therefore, high-quality approaches are necessary for network administration reasons, including agents able to interactively exploit these databases and give alerts.

As we have seen earlier in connection with the GreenSpace project, agents will also be needed in significant numbers to aid the end-user at the level of his workstation. The 3-D colour visualization of multimedia data streams makes this task more complex because some of its component parts find themselves at the frontier of the current state of the art. But, once again, the job is doable.

Active participation in a virtual-reality setting will also need terminals capable of handling six dimensions. Specifically, dimensioned for representation in 3-D, plus position, orientation and zooming (Chorafas and Steinmann, 1995). We will take a closer look at these issues later in this chapter.

METAPHORS, GREENSPACE AND THE WIDE AREA NETWORK INFRASTRUCTURE

A number of research projects involved in imaginative approaches to multimedia suggest that while the challenging tasks are many, one of the most vital is that of refining the *spatial metaphor*. Today's user-interfaces typically employ a rather naive desktop metaphor with all applications running on the system launched from this common reference point.

- In the future, virtual operating systems will necessarily need a similar concept, but in this case it will be a parametric spatial metaphor.
- At the same time, many of the 2-D concepts have to be converted to 3-D from the desktop metaphor, and many new ones have to be invented.

Any implementation solutions will have to account for the fact that virtual networks aiming to provide meeting places – like GreenSpace – as well as a foundation for cooperative work on a great many different specific tasks will feature a significant level of complexity. While each of the applications will need a design and feel that best represents its mission, it will be necessary to carry some common features consistently across applications.

In the case of GreenSpace, for example, when a given user enters the system there is a set of basic tools immediately available that will stay with him or her throughout the session. Such tools include interfaces to the method of navigation but also:

- Personal configuration editor(s),
- Interactive 3-D graphics,
- Easy-to-use means for realtime simulation, and
- A choice of teleconferencing approaches.

Valid solutions need to take advantage of computer and software advancements in combination with multimedia distribution technologies, such as the multimedia servers we studied in Chapter 6, and compression algorithms. A rapidly evolving wide area network (WAN) infrastructure is another 'must'.

As project GreenSpace aptly suggests, the wide-area characteristics of a virtual network will necessarily bring together disparate fields of voice and data communications – making use of whatever networking infrastructure is in place, but also extending its reach and functionality. Interactive sessions will be served through agents and include initiation, configuration, process and bandwidth management, as well as session termination.

Figure 8.4 Telecommunications, office automation and building automation correlate – with artificial intelligence as the focal point

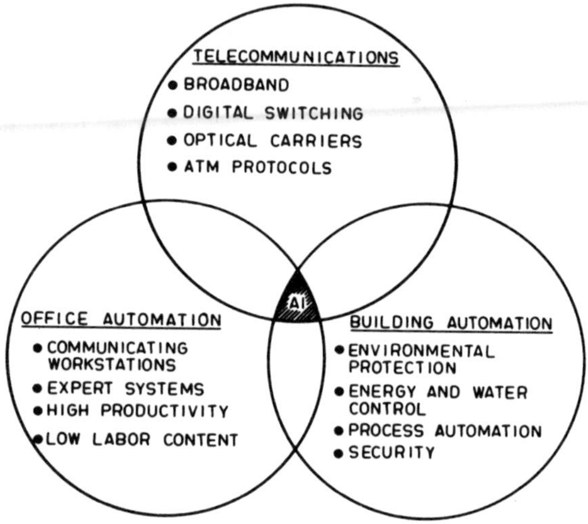

The basic concept is to get the network to act like an experienced human manager. It should appear transparent to the end-users except in well-defined and expected instances where agents step in to aid each participant in performing collaborative tasks.

Let us also keep in mind the synergy which exists between network structures and functionalities which, in the past, were taken to be quite distinct from one another. An example is given in Figure 8.4 which suggests that telecommunications, office automation and building automation today overlap in many areas – with artificial intelligence (AI) artefacts at the kernel.

When they are knowledge-enriched, metaphors can be instrumental in helping the network to automatically adapt to user requirements. For instance, in the cases of latency, burstiness, dynamic range, frame rate, data compression, and bandwidth-on-demand. Also in shared memories, level of detail, robustness, topology, scalability, manageability and protocols.

Protocols will need to be optimized for the types of transmission used for the short, medium-range and long haul. Both new protocols and new software will be born, as solutions are implemented and stepwise refinements are made towards the realization of virtual-reality systems.

GreenSpace suggests that flow control is necessary because it is unavoidable that some workstations as well as multimedia servers will have faster data transfer than others. The rate at which the network can send data to the station has to be negotiated either when the session is initiated or through configuration of the mechanism connecting the station to the network.

Within this context, during the setup phase of a session, the agent(s) helping each end-user can send descriptions of all objects that he will bring up to the session. GreenSpace, for instance, foresees not only audio and video but also tactile feedback.

- Solutions must address the correct selection or grasping of objects in virtual space,
- This approach will be provided by a series of tactors placed on or near the skin.

Objects in a virtual world may be given certain textural attributes which cause different sensations to the participants when they interact with them. This strategy can provide an important feedback channel in some situations, such as computer-aided design (CAD) applications where different materials may have different textures depending upon their strength, cost and other criteria.

It is already known from different studies that tactile feedback may be implemented in a number of ways, such as through pressure and vibration. Another issue on which GreenSpace is experimenting is *haptic feedback*. It gives an impression of weight and momentum forces and can be effective in producing a feeling of presence or satisfying interaction with objects in virtual space. The same is true of motion sensing, including:

- Procedural dynamics, and
- Parametric control.

Position and orientation tracking is a basic characteristic of a virtual- environment simulation as users interact with the world by moving their hands, arms or whole body. Positions and orientations can be tracked using a number of techniques, each with its own accuracy, resolution and range.

In the way the term is used with GreenSpace, *procedural dynamics* relates to the fact that certain realspace simulations may need to have associated to them natural physical phenomena such as gravity. Others may need to exhibit different behaviours when collisions are detected. Such dynamics may be encoded into the objects populating a virtual-space environment.

An example of parametric control is given by compound effectors. This is a mapping of many effects in a virtual space to a simple action on the part of the user. If the participant is managing a virtual earth-moving equipment as interface to a real earth-moving equipment (Figure 7.4 was an example), he may simply manipulate the virtual environment through gestures that cause many changes in the simulation to be mapped into the real world.

Since these virtual-to-real environment links may become complex and expensive, one of the major challenges of GreenSpace and similar projects is *scalability* in the number of simultaneous participants in a session. There is a need for optimizing the experience of users while, at the same time, providing an effective coordination.

NEW TERMINALS FOR MULTIMEDIA APPLICATIONS

To deliver the functionality which they promise, projects like GreenSpace require new types of terminals. These new terminals are evolving mainly in two directions. One is towards small-sized, lightweight, location-free capabilities aimed to allow users to talk with:

- Anybody,
- Anywhere,
- At any time.

The other direction is towards more complex intelligent terminals with multimedia features bringing moving image into conversations between persons, through a communication network. This approach also aims to present a large amount of processed and edited information in a form most easily understandable to users.

Both transitions towards new terminal design focus on location-free equipment and software. The change in concepts takes place as mobile communications are evolving from analogue to digital format. The goal is that of obtaining multimedia solutions with the ability to transmit, receive and massage data streams. But:

- *If* the ultimate goal in communications is terminals which allow immediate communication with anybody, anywhere and at any time;
- *Then* solutions will have to be fairly sophisticated going beyond what set-top boxes can offer.

Many problems have to be solved in communications using multiple media, including how a session is to be formed for each of the media, and how the individual media are to be multiplexed. In a challenge to TV-top boxes, in mid-June 1995 Packard Bell Electronics said it would take the home personal computer a step closer to televisions and stereos by providing a remote control.

The remotely controlled device is able to start the computer's built-in speaker phone, FM radio and TV tuner. It will also make it feasible to program the compact disk read-only memory drive, for instance to play a music CD. In this version, a thumbpad controls the on-screen cursor for running computer programs, but agents will eventually do a better job.

In other terms, set-top boxes on dumb TV sets is one approach to the creation of a multimedia terminal. Another and better solution is interfacing through personal computers. 'For interactive applications, the personal computer is going to win', says Nicholas Negroponte director of the Media Laboratory at the Massachusetts Institute of Technology *(International Herald Tribune*, 1994).

The main argument of companies betting on television set-top boxes as the best way into the home, is that almost everyone has a television and nearly anyone is more comfortable using it than using a computer. The counterargument is that rapid advances in computers are making them:

- Even less expensive than topboxes,
- Much easier to operate, and
- Much more useful as a terminal.

The newest generation of personal computers have chips powerful enough to run video with the resolution and stability of a television picture. Therefore, many communications, computer and entertainment companies are reversing course, investing in interactive services that can be used in connection to the personal computer instead of being television-centred.

There is another major advantage in this switch and it consists of the fact that once the problems connected to computer interfacing have been solved, the system can be employed in the office not just for entertainment. As I have already underlined, the office holds a much greater market potential than the home in terms of multimedia implementation – while the *virtual office* will push towards a merger of:

- Working in the business, and
- Working in the home.

No wonder, therefore, that Tele-Communications, the largest cable operator in America, has joined with Microsoft, the largest software company, to test programming products that allow a personal computer to access databases over TCI's cable network:

- Personal computer users might tap into the cable network on a business-oriented subscription basis, or
- Focus on interactive television programmes that permit the viewer to change the outcome as the show progresses.

As we have seen in Chapter 1 and in Chapter 5, joint efforts are being set by computer, software and entertainment companies that have already recognized the potential of the PC for interactive services and are rushing to capitalize on it. Intel has introduced a modem that hooks a personal computer to cable-television lines and allows PC users to receive information from the cable network 2000 times faster than through a conventional telephone modem.

As personal computers will increasingly offer video of television quality, many software publishers expect to announce entertainment and educational products to be delivered over cable networks and phone lines to PC users. Such projects tie-in nicely with the virtual-common concept of GreenSpace, since there will always be need for intelligent terminals to exploit the virtual-network capabilities.

THE MULTIMEDIA DATABASE CHALLENGE FOR CORPORATIONS AND FOR CONSUMERS

The focal point of corporate databases has been their utility in a market place which is transactional and decision-support oriented –

and where applications environments have steadily become more competitive and more demanding. Emphasis is placed on applications which require evolving high-power database systems, with the aim of improving the gains obtained from distributed databases in terms of:

- Service quality,
- Sophistication of applications,
- Cost reduction, and
- Greater productivity.

To reach this goal, both object-oriented and knowledge-enriched database solutions have been implemented successively overcoming a number of the constraints inherent in the implementation of information technology.

During the early to mid-1990s, one of the main features of database research has been the handling of long transactions and complex queries, as well as their implications in database management (see Chorafas and Steinmann, 1994a). Attention has been paid to the contributions offered by new database technology and the ways in which it can be exploited to provide tangible benefits to corporate users, including:

- Effective administration of database-mining operations, to improve business leadership,
- Exploitation of object-oriented approaches to enhance system development, and
- Use of memory-based reasoning (MBR) in connection to intelligent database solutions.

Methods have been brought forward for a balanced, effective approach to database implementation with the ability to handle requirements for fuzzy queries as expressed in the ISO Office Document Architecture and other, similar projects. Associated to this approach, as well as to the handling of long transactions has been the concept of distributed deductive databases, which we examine in the final section of this chapter.

By contrast, in the mid-to late 1990s, a focal point of interest is the development and use of multimedia database management systems (DBMS) to handle digitized video. While, as we saw in Chapter 6,

Oracle projects that the multimedia database management systems solution will involve specialized network servers, Microsoft sees a different future: networks of microcomputer-based multimedia DBMS will be the dominant force. In fact both vendors may be right.

- The sophisticated multimedia server will appeal to corporate databases which are more demanding in terms of functional performance.
- The microcomputer-based multimedia databases would most likely prove more cost-effective in the consumer market.

However, the priorities being put forward may be wrong. In 1994, Oracle was saying its information superhighway business might generate as much profit as its DBMS business within a decade. But revenues from information superhighway deals are still so scarce that Wall Street analysts do not even factor them into their mostly bullish projections.

- There is no multimedia business yet on the information superhighway, which would permit winners and losers to be distinguished.

At Oracle, the digitized video projects with telcos started back in mid-1993, when a group of young engineers worked on a program for sending video clips across a network. One of them got the idea of trying the program on as many as 1000 processors at once. But:

- While broadcasting a digitized film is a fairly easy matter, serving up 100 different films to 100 houses at 100 different times is a major scheduling challenge.

As we saw in Chapter 6, the concept has been to use a hypercube computer from nCube. The process is to split the video data into 1000 different segments and store each of them on a different disk drive controlled by its own microprocessor. One processor coordinates all the others so that the film flows seamlessly in the correct order.

On paper, this process can see to it that when a new viewer orders a film – whether it is 20 seconds or 20 minutes later – the first drives can start delivering the film. That is the power parallel processors provide. Something similar could be needed in a business environment for decision-support and other purposes.

In this sense, scheduling could be much less of a nightmare because parallelism might be used to take care of diversity. Anyway, in business and industrial applications management reports do not have the length of a full feature film, and there will not be requests by 100 different offices or meetings at 100 different times. There may however be other challenges connected to multimedia corporate databases, such as:

- Database mining, and
- Severe security problems (discussed in the following section).

Following database mining, interactive animation can effectively assist cognitive processing by simplifying the perceptual aspects of a reporting structure. As we have seen in Chapter 7, virtual reality can be employed to maximize the effective use of a visualization process for engineering and management reasons.

Interactive animation and database mining have some issues in common. Experimentation currently done looks at interactive animation as a paradigm for information retrieval, oriented towards the amplification of database mining. Such work is based on the analysis of several aspects of information, leading certain projects to the restructuring of the information retrieval processes within an engineering and managerial workspace.

CRUCIAL ISSUES ASSOCIATED WITH MULTIMEDIA DATABASE SECURITY

Security has always been a major concern with computers, communications and databases. It is no less true that the level of security that should be included in an information system involves considerable judgment about the dangers associated with security breaches – and therefore studies are necessary on ways and means for minimizing those dangers.

Database security has become a focal point because it plays an important role in overall multimedia information systems solutions. A number of parameters must be taken into account at the level of database design, implementation and maintenance, which have not necessarily been present with simpler data-oriented approaches.

- Part of the problem lies in the fact that today, rather than being applications-driven, the whole area of multimedia continues to be technology-driven.
- This change in hardware and software has evident aftermaths on the security solutions being provided, and the specific mechanisms supporting them.

Contrary to a technology-oriented solution, an applications-driven approach would focus on a requirements-analysis of the multimedia database, determining in detail both its users and the needs of each one of them in terms of security and protection. This goes beyond particular user characteristics and how one media relates to the other media.

For security purposes the primary and secondary locations of each user have to be ascertained; the clearance levels of each user-role; as well as the sensitivity and volume of each multimedia query or transaction. Detailed access requirements to the multimedia database must be outlined – and inference dependencies must be defined between media types.

Techniques for the arrangement of multimedia objects may be static or dynamic. The latter is preferable, but static solutions dominate today. As we have seen in Chapter 6, in most hypertext or hypermedia approaches:

- With current software the user may be browsing information by which related objects appear non-overlapped in different windows, based on his request.
- Tomorrow, however, the user will need dynamic approaches with temporal relationships and synchronization operators applied to the objects – not only from a query evaluation viewpoint but also in terms of security.

A conceptual model is necessary for assigning sensitivity labels for multilevel media access, granting users the privileges necessary for their functional role, but also permitting a fully distributed multimedia database design. Within an environment of deductive databases, agents can play a key role in this connection – as we will see in the next section.

Dynamic solutions can better serve security requirements by permitting the abstraction of the physical and logical details of the target multimedia database system. This is important both functionally and technically inasmuch as:

- Multimedia databases are the result of the integration of so far distinct technologies of databases and multimedia processing, and
- The right level of integration poses challenges which, if handled in an able manner, can offer innovative products.

It is self-evident that any solution regarding multimedia database security must pay attention to retrieval mechanisms not just for static media, like text and graphics, but also for media which are time-variant, such as animation – and more generally for the appropriate integration of digital audio and video streams.

The opportunity for doing a neat security job is present since all stored media are indexable by physical and logical attributes – even if, so far, only text and graphics have been indexable to a finer level of granularity by content.

- One possible solution that has been discussed is the *anchor* concept common in hypermedia models.
- It could be extended towards a more general mechanism, applicable to a wider range of dynamic media.

Parameters could, for instance, be included prerequisites connected to online information retrieval. Retrieval based on semantics is an important aspect of the manipulation of information in multimedia databases – as it is in all other cases of stored information whether in magnetic or optical disks. However,

- While significant progress has been made in image manipulation including zooming, rotation, smoothing and enhancement;
- Corresponding progress in secure extraction and presentation of the key features is slow.

Security controls should accompany user interactivity all the way to visualization, from panning (moving across) to zooming-in (widening), zooming-out (narrowing) and other operations. As the end-user is in control of the system, the viewpoint can be manipulated thereby altering the user's topic focus.

A pan changes the viewpoint on a resource without any of the scaling effects inherent in a zoom technique. In zooming-out, the frame of the media resource is expanded, so less detail is accessible but more of the resource is presented. By contrast, in zoom-in the frame of the media resource is reduced, giving more detail. Security risks may exist with every one of these operations.

Similar references can be made regarding the manipulation of spatial data on which rely many multimedia applications. Spatial data can be stored in different formats, from raster (bitmap) to vector format and video format. All of them need to be protected.

In bitmap format, an image is stored as a 2-dimensional grid of points. In vector format, an image is stored as a collection of shapes at specific locations. The vector image is usually created through inter active selection and positioning of shapes by the user. The video signal format is most suited for the storage of moving pictures and animation.

Each format may contain original, raw data as well as manipulated information. Such manipulation may be *legal* or *illegal* – but subsequently taken as being legal, once it resides in the database. A comparable concept is valid about other information elements and their usage, such as:

- Textual description of the image contents,
- Colour maps for colour images, or
- Enhancement history of the image.

In conclusion, the study of security measures should account for the fact that multimedia objects have various dimensions which are parametrically accessible, each of different duration. Hence, the solution which we choose will have to deal with the integration of all retrieved objects of different media types in a synchronized and secure way. Attention to security is necessary from the overall system design to its specific implementation.

DEDUCTIVE DATABASES, INTENTIONAL AND EXTENSIONAL SOLUTIONS

Deductive databases constitute a relatively recent research area and reflect a growing awareness of the needs of using knowledge-enriched tools for design, representation and control. In the background of this approach are the benefits to be derived by using logic as a means for the formalization of both the static and the dynamic aspects of multimedia databases, the way they have been discussed in the previous section.

Whether enriched by intelligence or dumb, a database system is usually organized according to a data model, and it aims to provide a formal framework for the representation and manipulation of information elements.

- In past years, dumb solutions dominated – but today they are totally unsatisfactory as database requirements steadily increase.
- Because, however, we cannot throw away all legacy databases, the solution has been advanced of *intentional* (hence deductive) and *extensional* approaches.

Applications reasons justify this change. Intentional databases are selective, knowledge-enriched and operate by means of expert-systems rules, through fuzzy logic or other algorithms. Typically, database intelligence resides at the workstation(s) and at the server(s) – and it works in an interactive sense (Chorafas, 1994a).

Extensional databases are massive, representing vast amounts of information, usually data, stored over decades of computer practice. Among DBMS, those object-oriented have *intentional* features. By contrast, the three earlier models for database management: hierarchical, networking (Codasyl) and relational models are extensional.

Based on a client–server model, Figure 8.5 explains the difference between these two terms. As an inference engine, the workstation acts as an intentional database (IDB) search mechanism. The actuators may be agents or other knowledge-enriched artefacts (see for example Chorafas, 1994a).

Knowledge artefacts also reside at the server where the extentional database (EDB) is. In this case, IDB is a layer of intelligence over EDB – which in old times was known as *Logic over Data*. Together the two IDB: one at the workstation's database, the other at the server, constitute the network's knowledge-bank management system (KBMS).

This amount of intelligence embedded in a system setting is necessary for a number of operational reasons. Not the least is the fact that new applications are interactive, and have to make use of knowledge-banks. The level of required sophistication has simply changed. New applications also involve complex operations:

Figure 8.5 Parallelism in networked inference engines and databases

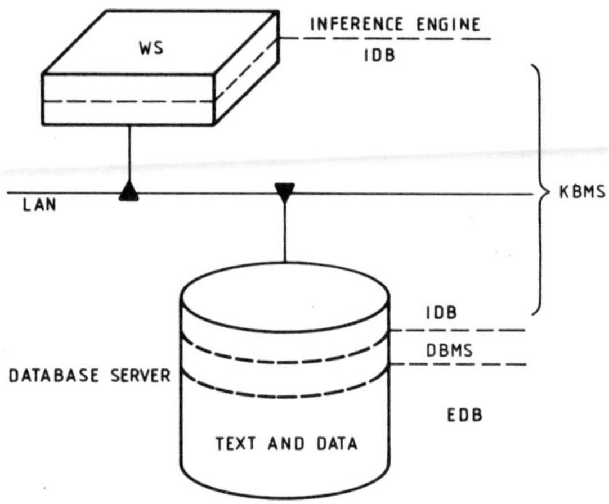

- A design database requires the support of composite objects and different versions of the same object.
- A multimedia database may contain variable-length text, graphics, images, audio and video information.
- A knowledge-based system needs information rich in semantics, as well as heuristics able to handle vagueness and uncertainty.

Hierarchical models are too inflexible to work in this evolving environment, and therefore preclude advanced applications. Yet, an *ephemeral hierarchy* is important for semantic reasons. This duality of having and not having a hierarchical structure is supported through object orientation (see also Chorafas and Steinmann, 1993a).

While the relational model with its flat files presented significant advantages over the hierarchical and the Codasyl models, it has the disadvantage of the lack of semantics. To compensate, the relational model imposes restrictions on the representation of data.

- By organizing information elements into tables it presumes horizontal and vertical homogeneity in the data which is far from being a real-life situation.

- This sort of treatment is incompatible with multimedia databases, where the variable-length data structure of images, graphs, text and voice cannot fit well into the relational framework.

The new generation of applications no longer deal with fixed-format records, but require access to multimedia information elements on the basis of the structure of a graphical item or by following logical links. Relational (and naive) query languages, such as SQL, were not designed for such applications.

The fact that relational models do not explicitly include semantics as part of data representation, leaving it to application programs to interpret data semantics constrains their usage. At the same time, operations in a knowledge-bank are much more involved than those in a traditional database. When a rule is added, the system must check for contradiction and redundancy, a process known as *equilibration* (see Chorafas, 1994).

- Such operations cannot be represented directly by relational tables, and
- The complexity of checking increases rapidly as the size of the knowledge-bank grows.

Advanced applications require the representation of complex data elements as well as dynamic relationships among them. It is therefore not surprising that users in multimedia environments have found relational technology inadequate in terms of flexibility, modelling power, efficiency and usability.

These characteristics do not escape the attention of computer vendors. But they are faced with a rather severe conflict of interest: between renewing their systems and serving the real needs of their clients on the one hand, and on the other losing the cash-cow of steady income from obsolete hardware and software installations. Therefore, user organizations who are serious about multimedia implementations must be very careful in their choice. Not everything that glitters is gold.

Part Three

Projects on High-Performance Computers and Networks

9 The High Performance Communications and Computing (HPCC) Programme in America and its Challenges

INTRODUCTION

In spite of mergers, acquisitions and alliances in the media and the telecommunications industry – which we saw in Chapter 1 and followed up in several other chapters – huge innovative projects like those we have been discussing cannot be financed solely by private industry. There is also the important issue of training the technologists needed to undertake these projects. Both are two good reasons for government support; hence, for taxpayers' money.

Fifty years ago, Dr Vannevar Bush, of MIT, promulgated the vision of federal support for university research. Assisted by the GI Bill, for World War II veterans, this strategy implicitly extended Federal support for the education of undergraduate and graduate students – as well as for a significant amount of projects which have been sponsored by the different agencies of the US government at American universities.

Over the last 50 years, two of the better-known agencies of the US government who have been heavy sponsors of university research projects are the National Science Foundation (NSF) and the Advanced Research Project Agency (ARPA) of the Department of Defense. Both were instrumental in bringing forward projects and programmes which without solid financing would never have made it.

The latest major government-financed project, and one with relevance to the subject of this book, is the High-Performance Communications and Computing (HPCC) programme. It is sponsored by a number of blue-ribbon US government agencies:

- Advanced Research Project Agency;
- National Security Agency (NSA);
- National Science Foundation;
- National Institute of Science and Technology (NIST);
- National Oceanic and Atmospheric Administration (NOAA);
- Department of Energy (DOE);
- National Aeronautics and Space Agency (NASA);
- Environmental Protection Agency (EPA);
- National Institute of Health (NIH); and
- National Library of Medicine (NLM).

While HPCC is not exactly multimedia, its support for advanced communications solutions and for parallel computers provides the necessary infrastructure for the type of applications we have seen in Part One and Part Two. There is also a parallel to HPCC, known as High-Performance Computers and Networks (HPCN), sponsored by the European Union.

Hence a study of the HPCC project is important because of the lessons which can be learned on what went right and what went wrong. Also because of the aftermaths which are very recently coming under perspective in America, and which most likely will have a very significant impact on European governments in regard to sponsored research.

GOVERNMENT-SPONSORED RESEARCH AND HPCC

The Vannevar Bush contract, of which we spoke in the Introduction, is now being questioned by the US Congress, while academia and many business leaders strongly believe that reducing funding for information technology research at US universities would be counterproductive. But times have changed and it is wise to take notice of this change. While both:

- The university infrastructure, and
- The US computer industry

rely on university research projects supported by the National Science Foundation, the Department of Defense, and other agencies to construct the foundation from which to build future technology, the rationale of this sponsorship is being challenged. In fact, the first major challenge to Dr Bush's concept did not come in 1995, or from the US Congress – but in the late 1980s when, in the UK, Margaret Thatcher killed project Alvey and other government-sponsored advanced research programmes.

One reason for this change in policy is the growing budgetary constraints. Another reason is that pre-competitive research lacks clear goals sustained by well-coordinated, market-oriented programmes in high-technology.

American universities and American industry object to reducing funds for the High-Performance Computing and Communications initiative and other National Information Infrastructure (NII) related research. It is thought that budget cuts would seriously jeopardize the progress the researchers have been making in the critical areas of computing, communications and software. But are these achievements real or imaginary? This is what we examine in Part Three.

Which are the project areas related to HPCC and NII for which American universities and American industry would like to see government financing continued – if not increased in a significant manner? While a complete answer to this query will lead to a very long list, the top ten research domains are as follows:

1. Agile end-user interfaces, interactive visualization, vision and recognition;
2. Agents and other knowledge artefacts;
3. Very-high-level programming languages and visual programming;
4. High-performance distributed computing;
5. Distributed multimedia databases and image transfer;
6. Intelligent broadband networks and associated protocols;
7. Human genome and molecular genetics;
8. Bioengineering and new pharmaceutical designs;
9. Quantum chromodynamics and other advanced projects in physics;
10. 72-hour weather prediction and pollution dispersion.

A well-financed, government-sponsored programme along this frame of reference would not only provide a far-sighted vision of the underlying technologies, but also give a major boost to the competitiveness of American industry. Its critics, however, suggest that many of the current projects lack focus and therefore cannot serve major competitive needs. Rather, they are spread thin over wide areas, or reinventing breakthroughs of the past.

- The fact is that nothing practical has come out of HPCC over the years of its existence, and nothing practical is in sight.
- With the momentum Internet has today, it is very difficult to come up with something competitive when rusty ideas dominate.

Since the primary goal of HPCC is to promote research in universities and to assist graduate studies, specific criteria should be established to judge the outcome. The state of Bavaria, in Germany, has established parameters for funding university research projects and universities at large. These are:

- Number of graduating students at the university, and
- Number of Nobel prize-winners teaching or doing research at the University.

The first criterion keeps students on the run to finish their studies in order to continue receiving government funding. The second, has to do with faculty excellence. There are three Nobel prize winners, for instance, at the Technical University of Munich.

Both free-wheeling financing and the establishment of criteria of excellence pose a major dilemmas. On one side, information technology is an integral part of a survival strategy. At stake is the ability to interrelate and interconnect all key components of corporate policy. There are also the *grand challenge* problems – from the human genome and nonlinear models for derivative financial products, to interactive multimedia for corporate and consumer uses.

On the other side, government-sponsored pre-competitive research in colleges and universities is not easy to quantify in return-on-investment terms. Now that the Cold War has ended, the criteria have changed. And legislators are under steady pressure to reduce the budget deficit. This is true not only in America, but in all First World countries.

One of the stumbling blocks is the lack of perspective. Grand challenge problems should not be viewed under the narrower viewpoint of the physical sciences, though there exists in the natural sciences a great deal of very worthwhile research to do. Every one of the tough problems facing science, industry and business at this time, requires a growing amount of new technologies. For instance,

- From *giga* instructions per second (GIPS),
- To *tera* level (one trillion operations per second) through massively-parallel computing solutions.

In America, this is what the High-Performance Communications and Computing (HPCC) programme aims to provide – and, in the process, bring the United States, its laboratories, its companies and its people, into the 21st century.

Or at least this is what was said in 1991 when officially the programme started. A few years down the line it is appropriate to ask: 'Have these goals been attained so far?' If not, what kind of lessons can we learn to avoid repeating the same mistakes all over again?

This is a question asked today by lots of cognizant people. It is always rewarding to take a major programme into perspective, examine its achievements, highlight its uncertainties, and identify its failures – then drawing factual and documented conclusions.

The more critical is this evaluation, the better it is for all concerned. Therefore, it should be appreciated that this discussion is not intended to be simply critical. Rather, its purpose is to be constructive by placing emphasis on what has been done and what should be done in order to obtain commentable results – and sustain government financing.

THE GROWING CRITICISM OF HARD SCIENCE AND BIG SCIENCE

The rethinking of priorities for government financing reasons does not only come from politicians and outsiders. By mid-1995, senior researchers in computers, communications and software have been suggesting that it is necessary to find a new engine, different from the High-Performance Computers and Communications programme.

The reasons for this change in heart are more or less clear. While advancing under the HPCC flag, research on computers and communications became a visible component of US federal science and technology policy – but not all projects reached the expected results.

- Budgets grew enormously, as did the influence of Washington's computing research bureaucracy.

- Snowed under by the workload, the curve of imaginative advances bends, and hence the current research, development and implementation (R,D&I) engine runs out of gas.

Other issues also start making the headlines. The sense that research fields need big conceptual cores seems to stem, at least partly, from the observation that the *hard sciences* such as physics and astronomy have been successful in gaining substantial government support for major experimental programmes – involving accelerators, telescopes, gravity-wave detectors and so on – but this view, also, is now challenged.

A growing number of scientists now believe that the sense of the broader value of *big science*, while probably having a grain of truth, seems to be overstated. For one thing, a sizeable chunk of the big science money goes to engineering and construction firms, not to research projects. Furthermore, as the scientific project gets under way and gains momentum,

- Many begin to devour funds,
- While results come very slowly.

For instance, senior research physicists have recently been concerned that projects at the level of the supercollider would be built at the expense of support for smaller, single-investigator research projects. Some biologists are of the opinion that the Human Genome project would elbow out very important small research projects that needed to be done on molecular genetics.

Similar arguments are being made about HPCC and the big science computers and communications projects at large – including the information superhighway. Hence the drive starting in the communications and computing research community for a new concept to replace the HPCC initiative.

As it is to be expected, this opinion is not universal. A switch in funding evidently touches established interests and leaves some projects orphaned. Therefore, other people argue that HPCC does not have some of the downsides of big science.

- Although funding amounts to a visibly large amount of money, with over $1 billion annual expenditures,
- The argument is that this programme is not structured as one

large, single project, but rests on many smaller projects scattered over different agencies.

This makes the management of HPCC a little more diffuse, and the politics of what should and should not be done more complex. But is HPCC properly focused? Can it help develop and get full potential out of a new National Information Infrastructure (NII), a notion which at Capitol Hill has broad bipartisan appeal?

Still more to the point are there any lessons which could be learned to influence similar initiatives in England, Germany, France and other European countries – as well as in Japan? The best way to respond to this query is to take a careful look at high-performance computing and communications projects, including their aftermaths.

HIGH PERFORMANCE COMPUTING PROJECTS AND THEIR AFTERMATHS

Sponsored by the Office of the President of the United States, HPCC has been propelled by the recognition that high-performance computers and communications networks are becoming increasingly important to advancements in science, industry and finance. Based on this concept, a positive answer was given to the query posed in the last paragraphs of the previous section – in a 1995 report by the Computer Science and Telecommunications Board of the National Research Council.

This report, titled 'Evolving the High-Performance Computing and Communications Initiative to Support the Nation's Information Infrastructure' was prepared by a panel chaired by Dr Fred Brooks and Dr Ivan Sutherland – both authorities in the computer field. The Brooks–Sutherland report, however, makes a much broader argument than HPCC. And there exactly lies the heart of the matter.

The thesis advanced by Brooks, Sutherland and many other scientists is that research projects in communications and information technology are a moving gear of the economy. But they have to be instrumental in global economic competition for products and markets:

- Technology is reaching the point of having a transforming effect on our society, our industries and our educational institutions.
- The goal to accelerate in a pronounced way the availability and

utilization of the *next generation* tools is indivisible from that of future competitiveness.

- Survival requires a polyvalent but well-tuned research, development and implementation (RD&I) effort, without stiffening the arteries of innovation through tunnel vision.

Notice that emphasis is placed not only on R&D but also on *implementation*. Too many research projects are born, live and die in the laboratory – or their fruits take so long to be brought to the market that they end by spoiling their market opportunity.

The changes taking place in our economy and in our society are significant, but the process of change accelerates and therefore *time-to-market* is an increasingly crucial issue. Just as important is to establish a sense of direction. 'I feel the real thing in this world is not so much where we stand, as in what direction we are moving', said at the beginning of this century by Oliver Wendell Holmes (Chief Justice, US Supreme Court).

In 1991 the Office of the President forecast that in terms of computer power and its practical usage the teraops level will be reached by the mid-1990s. This has not yet been achieved in meaningful economic terms but the trend is shown in Figure 9.1, which presents a scale of just 'note difference'. This term means that only order of magnitude matters. At the time, it was projected that by the mid-1990s

- Those scientific, industrial, commercial and financial organizations with a clear concept on how this power can be put to use will move ahead.
- The others, that is the laggards, will lose their competitive edge as well as their market standing – and once lost, competitive advantages are very difficult to regain.

How true and how damaging can this falling behind be in terms of competitiveness is easily documented by what has happened in 1992 to IBM, particularly in the October to December timeframe. As the stock dropped to $43½, Wall Street investment banks estimated that IBM's capitalization stood well below the value of its assets – essentially leaving a negative good will. (Since then, because of the change taking place in corporate culture and better management, IBM stock has rebounded to over $100.)

Figure 9.1 Computer power necessary for grand challenges can be provided by scalable parallel designs

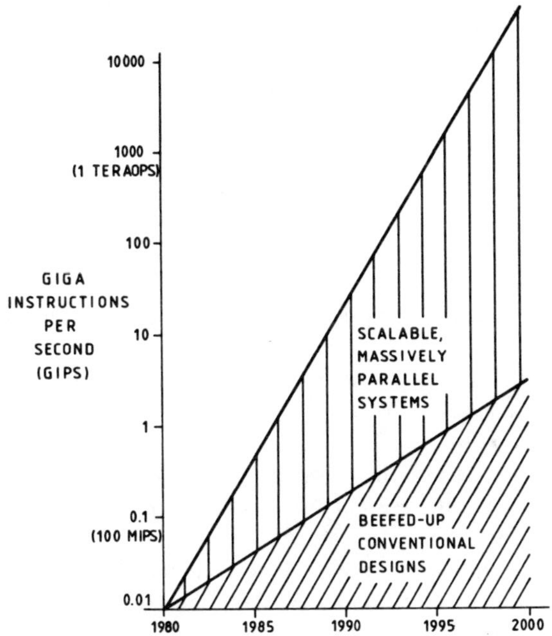

The market votes with dollars and its voting tends to reflect the downgrading of a company and its products. The study which went into the graph in Figure 9.1 exemplified the inability of beefed-up conventional designs to reach 1 gigaops by 1995, or to go beyond 5 gigaops when, by the year 2000, massively-parallel *ultracomputers* will have attained beyond 10 teraops (10 000 gigaops).

In this figure, as in others in this text, the fact that straight lines are being drawn is not necessarily precise. But since direction is the crucial element, the intention is to help the reader *notice the difference* in the sense of the most likely trend.

- Many projections on future trends happen to be conveniently drawn as straight lines.
- Yet, we all know that the data points on computer performance are far from linear with respect to time.

One of the reviewers of this text raised the query: 'What is a *beefed-up* conventional design?' While the term 'conventional' computer design is understood by context to mean architectures based on a single processing element, a typical cluster or a vector processor, there is no common ground in explaining what 'beefed-up' means.

As the reviewer aptly suggested, one can say with confidence that asking 100 experts with a basic understanding of computer system architecture to define 'beefed-up conventional design' would result in about 100 distinct answers.

- What is it that is being significantly improved in these patched up solutions?
- Does this *beefing-up* apply also to the operating system and/or compiler technology?
- Or is the sought-out methodology only being applied in connection to hardware?

The latter is of course the simpler case, and in this sense it has been employed in the present text. An example is IBM's addition of a vector processor to its top-of-the-line 3090/600 computer. It did not change much in terms of performance, but it did have a very negative impact on the costs. Without the minimal doubt, beefing-up should not be part of a company's strategy.

Admittedly, IBM did not succeed in bringing out a vastly improved computer solution. But can it be that the HPCC programme has reached its goals? Seen under a 1995 perspective, the answer is not necessarily positive.

- The teraops speed has been reached, but the market has not been developed. As a result the cost of the teraops engine is way too high, which further limits market potential.
- The market shows apathy because on one side the military budgets have been cut – and on the other there is no software to exploit that high power.

Software is the Achilles heel of HPCC, as it is of so many other parallel-computer projects. The first lesson to be drawn from this government-sponsored programme is that too much attention has been paid to hardware and too little to the sophisticated software needs. The allocation of resources is unbalanced and, therefore, nobody should really wonder that the results are not that good.

EMPHASIZING THE ABILITY TO DELIVER

It is rather curious how some companies as well as people confuse stop-gap measures with solutions for the longer trend. Yet, a person's, a company's, and a nation's strategic plans should provide evidence of appreciating the difference and of recognizing the effects of technology and of markets in short, medium and long-term timetables.

In the particular case of HPCC, the case concerns the effects of properly exploiting an unprecedented computational power for competitive reasons. In the last analysis it is the deliverables, not only research, which make the difference between responding or not responding to grand-challenge problems.

- The insight a well-coordinated research, development and implementation approach can present is foremost to post-industrial *survival.*
- The concept behind this emphasis on fast delivery is new, and it has been explained as *time-to-market.*

Focused research and fast delivery are fundamental issues of the 1990s and the early part of the 21st Century. We simply cannot afford to forget about innovation, but innovation does not serve if our products do not hit the market ahead of the competitors.

Even if many companies and their projects still do not see deliverables to be a top-level priority, that's the way the market perceives and appreciates the return-on-investment in R&D. As we will examine in more detail in Chapter 11, projects which give no results should be killed even if they are considered to be vital. Then they should be restructured and reinstituted:

- With better focused objectives, and
- Under new management. (That is the strategy Bankers Trust is following with its information technology projects. And the results are very rewarding.)

To properly appreciate the message conveyed by the foregoing paragraphs, we should keep under perspective that the 1990s have already seen a tremendous change in the way the market looks at the benefits to be derived from computers and communications. There has also been a major switch in the manner the vendors themselves conceive their survival in an industry that has become more competitive than ever. Two points are outstanding in this connection.

1. The long-running debate about how best to make computers thousands of times more powerful than they are now, appears to be ending by emphasizing the wrong parameters.
2. Curiously, no vendor has really been able to out-distance himself from the crowd and get positioned to exploit the forces of the mid- to late 1990s.

In terms of issue No. 1, as of 1995 the leading American computer designers seem to have agreed that parallel-computers technology has the most potential for the next generation of machinery. But at the same time they do not seem able to break out of the self-imposed limitations of old software – and the US government which sponsors these projects does not seem willing or able to push them out of it.

Massively-parallel processing solutions involve thousands or tens of thousands of independent processing elements teaming up as one computer to solve problems far too imposing for even the most powerful vector processors available today. But *if* we are going to move:

- From an era in which massively-parallel computers were thought to be the exception,
- To one where they have become a tool for scientists, engineers and financial experts,

then we have to break the software bottleneck. In the 1980s, software was the name of the game in gaining market leadership in computer sales, but by the 1990s it has curiously moved out of the top spot.

It is true that in a systems sense there are also other preoccupations. System integration in a heterogeneous environment, as most computer installations presently are, calls for a large amount of computer power which has to be delivered in a very cost-effective manner – but even more than that it needs sophisticated software.

The required level of *high efficiency* and *low cost* can only be reached by the new computer architectures, not by the old – if and when appropriate software becomes available. In fact, the most recent American forecast is that conventional supercomputers would soon be obsolete.

- Conventional supercomputers as well as large mainframes line up billions of calculations, then use one complex and expensive processor to race through them.
- By contrast, desk area networks and parallel computers divide a problem and distribute it among hundreds or thousands of simple, inexpensive processors.

Not only does this make feasible a better-balanced solution supporting much greater power at significantly lower cost, but also it presents a more rational systems architecture able to address current problems in an efficient manner. Mainframers, however, would argue that this argument is not yet proven.

As it cannot be repeated too often, mainframers wrongly advise their clients to attach their 120 MIPS workstations to 110 MIPS mainframes. That is totally ridiculous. What users need is to attach their high MIPS workstations to their 10 000 MIPS server or their 100 000 MIPS server. That makes sense – provided the software bottleneck is broken.

THE FIRST RESULTS OF THE HPCC PROGRAMME IN AMERICA

The goal of the US High-Performance Communications and Computing (HPCC) Programme has been to significantly accelerate the commercial availability of high-performance computers by promoting both development and effective utilization of *the next generation* of computers and networks. But, as we have seen, there are software snags.

On paper, if not in practice, the concepts behind HPCC are simple but far-sighted. The programme is built on the premise that recent advances in high technology offer the potential for:

- A 1000-fold improvement in useful computing power, and
- A 100-fold or greater performance in available computer communications capability, including database exchanges.

An impressive statement made in 1990–91 was that both goals were considered to be doable in five years' time. The experience of the five years which have elapsed proves that this was hollow. The challenge is to define in which way goals have not been met – and why what was achieved in the interim was *not* put into practical, useful work.

Yet, practical implementation is precisely the mission HPCC should have put to itself. It is real-life applications, not sheer computing power, which lay the foundation for changes in competitive position at all levels of:

- Science,
- Industry,
- Finance,
- Education, and
- Government.

Attention to software should have been that much greater also, due to the fact that in the United States the HPCC Programme is a strategic federal investment in the frontiers of computer and communications technologies. It has been formulated to satisfy national and corporate needs from a variety of perspectives which include:

- Science applications,
- Business implementations,
- Effect on human resources,
- The overall process of transition from current to future systems, and
- Infrastructural technological developments.

From *millions* of instructions per second to *peta* instructions per second, the difference is 9 orders of magnitude. Neither human know-how nor applications change so fast to be able to capitalize on this colossal computer power – unless:

- We plan well in advance how to go from 'here' to 'there',
- Focus on those applications which stand to benefit the most from greater power, and
- Point to the right direction in terms of implementation as well as *skills*.

This right direction involves a significant *paradigm shift*. While up to 1990, weapons systems, nuclear engineering and aerospace were the clients of choice of very fast machines, the 'peace dividend' saw to it that this market shrank tremendously. Business, industry and finance is the new market – but the prerequisites are to:

- Develop visual, user-friendly and efficient parallel languages,
- Prepare for the implementation of the new environment through training,
- Make available to ourselves parallel operating systems, database management systems, and transaction processing monitors, and
- Be able to manage change all the way from research to implementation.

Proponents of easy-going government-financed projects would say that the *need* for greater power overrides all other issues. But they are wrong. If there is no *cultural* change, the extra power will be for the birds; it will not be put to profit.

The management of change is an issue which sounds like motherhood and apple pie. Everybody seems to be for it, at least on the surface, but important undercurrents work against it. People are afraid for their careers and often fail to appreciate that there is no future for those who live in the past.

GOING BEYOND PRECOMPETITIVE RESEARCH

Correctly, the focal point of the HPCC programme in America has been the underlying science, engineering and technology required to carry out the missions of the late 1990s and beyond. The fundamental issues being addressed cut across various disciplines – but also through a number of US government agencies. What they have in common is the fact that they address precompetitive research.

However, judging from the experience of other imaginative high-tech projects like MCC in the United States, Alvey in the United Kingdom, and ESPRIT in the European Community, *precompetitive research* does not reach the heart of the problem. Not just some strides in technology but also the fast commercialization of what is done makes the difference between success and failure.

Without appropriate software, commercialization of high-tech in computing machinery cannot take off. Relying only on drum-up sales is the wrong policy in connection to massively-parallel computers. The current vendors' way of pushing mainframes out and parallel computers in is taking the user organizations to the cleaners – because they rely too much on converting old applications to the new machines. (This also explains the wrong emphasis put on Fortran.)

- This is the worst possible way to exploit the effectiveness of the new computers.
- The prejudice it creates hits the vendor's own interests in the medium and longer term.

Beyond the trust issue between user organizations and their vendors is the realistic appraisal of the situation. Hardware technology has evolved tremendously, but advances in programming techniques are even more necessary than 100 000 MIPS in order to convince skeptics in the user community that a radical change is long overdue.

- Many problems were until now approached only by expensive bulky mainframes, characterized by the so-called von Neumann bottleneck, and therefore were very complex to analyze while some could not even be handled.
- Now these problems can be attacked at much faster speeds – and most importantly in a relatively simpler manner – by massively-parallel machines.
- But languages like parallel Fortran, a construct whose origin dates back to 1957, and IBM's 704 computer, are far from breaking the software bottleneck.

'It is not sure that a parallel program is simpler than a sequential one', said one of the reviewers. This observation has merits. As long as the vendors of parallel computers shy away from the much needed design of new languages and stick to old hats like Fortran 90, programming is not necessarily going to be simpler. In fact, it can be much more complicated and expensive.

Another reviewer reacted in a similar way: 'Having personally programmed over five different types of parallel machines, both commercial and experimental, I can say with confidence that there is nothing simple about solving problems using parallel machines.' That is not a universally held opinion, but the point is that *software sells hardware* – and the software is not available.

Visual programming enriched with knowledge engineering is a strategy which can help to capitalize on the incredible power of the microprocessor and the very fast developments in the architectural domain. But projects have to go well beyond precompetitive research into real life applications and productization.

In conclusion, there are a number of impressive developments in hardware and there are many more in the making. But years of intensive research done by the author in America, Japan and Western Europe failed to reveal similar breakthroughs in software. Let's hope that this unbalanced situation will not last for long.

SEEKING SOLUTIONS THROUGH HIGH-PERFORMANCE COMPUTING

In a study under the title 'Seeking Solutions Through High-Performance Computing', the US Congressional Office of Technology Assessment which is responsible directly to the Office of the President of the United States, calls high-performance technology 'The Future Tool of International Competition'. According to this top government agency, the challenge is to:

- Identify promising technologies and their most likely evolution,
- Establish directions for supercomputing infrastructures,
- Evaluate progress in high-performance computing and intelligent communications, and
- Make corrections in the course being followed, steadily focusing on measurable results.

Among topmost computer designers, one of the most interesting references made in connection to this strategy is that change takes place not in one but in three directions at the same time. As shown in Figure 9.2, this is a far cry from what is offered today by aged mainframes – which are dwarfed by the new generation of computing equipment.

Cognizant industry executives, versatile in the domain of computers and communications, underline that it is quite important to assure *how square* the solution is – along the axes of reference shown in Figure 9.2. Namely,

Figure 9.2 A frame for high-performance computing along three axes of reference

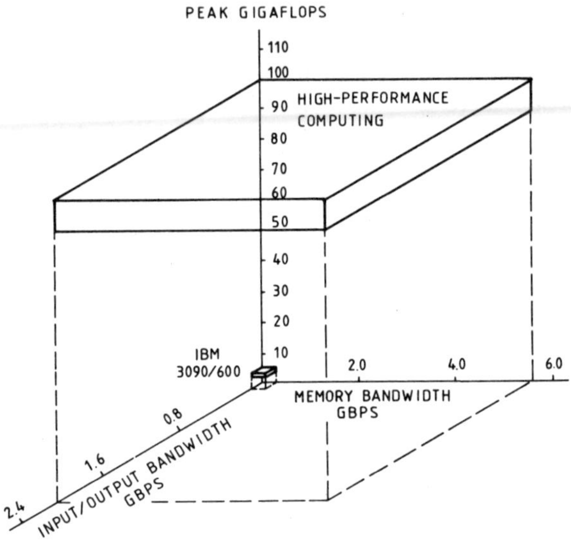

- Peak gigaflops,
- Input/output bandwidth, and
- Memory bandwidth.

The more symmetric the system which we obtain, the better balanced is the implementation perspective – which means the solution to which it contributes. A well-balanced approach works to everyone's advantage, from designers and vendors to users.

In fact, mainframes are not just suffering from very high cost per MIPS being delivered and from contention in terms of cycles and central memory due to the shared-everything architecture – or, in other terms, the tightly-knit mainframe-oriented hierarchical structures. They are also, if not primarily, limited by *their memory bandwidth*. This tends to *invalidate* the very expensive special-purpose processors they feature, even if their vendors keep silence about this important fact.

Why has a well-financed project aimed at assuring the future competitiveness of American industry paid no attention to the need for balance along the axes of reference shown in Figure 9.2? I don't pretend to know the exact answer, but my guess is that it is not independent from the fact that HPCC had many fathers – each looking after his own interests.

The HPCC programme is constituted of major component areas which represent key challenges of the 1990s. But is this work properly coordinated? And who is doing the integrative job? (How successful or unsuccessful this effort has been is discussed in Chapter 4.) Here are the areas in reference:

High-Performance Computing Systems (HPCS)

Emphasis is placed on the development of underlying technology for scalable fully-parallel computers capable of sustaining trillions of operations per second (teraops) on large-scale problems.

HPCS aims to produce scalable fine-grain parallel-computing systems through the development of prototypes able to *demonstrate* a 100- to 1000-fold increase of sustained computing capability before the end of this decade. This two-to-three orders of magnitude measure is in comparison to today's machines that follow the more conventional design path. Notice that the Japanese Real World Computing (RWC) programme has a similar goal.

The HPCS programme is executed under four sponsors: ARPA (Advanced Research Projects Agency of the US Department of Defense (DoD). ARPA has the largest dollar commitment of the four agencies), Department of Energy (DOE), National Science Foundation (NSF) and National Aeronautics and Space Administration (NASA). The four agencies share responsibility and lead the focus on technological challenges since the early stages of product development.

- Prototypes allow empirical evaluation of alternative solutions,
- The analysis and design of applications will be restructured as performance increases.
- Experimental results will be fed back into the design process to refine successive generations of systems.

Research on future generations of computing is expected to produce the underlying scaling concepts, components and packaging.

Architecturing will assure that the required advanced technologies are available for the implementation of new systems solutions, into which existing installations could be recast.

Advanced Software Technology and Algorithms (ASTA)

The goal is the development of generic software technology, algorithms and heuristics for grand-challenge research applications in a networked environment. The key role in this domain has been assigned to NASA. But is NASA in a position to answer what has already been said in terms of huge gaps in software?

- Making feasible new approaches to business, industrial and financial applications, not only in science and engineering, and
- Significantly improving user-friendliness, by enhancing software productivity beyond what current computer-aided design approaches (CAD and CASE) can offer.

The key to success in this effort is a thorough knowledge of the applications environment to enable the use of experience gained on leading-edge implementations in other fields, thus guiding future software development and maintenance efforts. For all its other accomplishments, NASA lacks experience in finance and in manufacturing.

We will not be concerned in this text with the parallel goals of HPCC: basic research, education, and human resources. These are very important issues and deserve lots of attention, but they are outside the scope of the research which led to this book. Instead, further emphasis will be placed on the applications perspectives as well as on the *real challengers* of HPCC, the tiny microprocessors.

ADOPTING A STRATEGY FOR PARALLEL COMPUTER USAGE

Over the years, major computer manufacturers favoured small stepwise improvements that protect their investments in old technology. That is why, as of recently, some of their best customers defected from them.

After having lost a huge chunk of market share, some of these vendors finally see that parallel processing today represents the only way to increase computer performance dramatically. But it is not easy to attain market penetration if other prerequisites are not met.

On the user side there is slowly developing a new trend in computer usage. An example is provided by the strategy followed by the foremost financial institutions and the leading manufacturing companies, which aims at *damage control*:

- Capitalizing on the long-developing schism between the more advanced computer user organizations and the industry that supplies them, and
- Going to the sources of that schism by exploiting the changing economics of the computers and communications business.

Before computer users vote with their budgets, they are asking two critical questions: Is the new technology truly better than the old? And if it is, what benchmarks should be used to figure out which of the new platforms is right for a given application?

Tightly costed, high-value-added and flexible computing solutions are competitive tools. This is particularly true as the manufacturing, financial and other industries see themselves as the low-cost producers of the products or services they offer – which has become a prerequisite to corporate survival.

Investment banks, for instance, have long understood that principle. They also appreciate the fact that in today's fiercely competitive markets no mature financial products provide significant profits. Risk and reward comes from innovation.

- As the derivatives business helps to demonstrate, anything can be re-invented in Finance.
- The key is RD&I through the application of advanced information technology (see also Chorafas, 1995a).

Solutions must not only be *innovative* but also come at a *fast pace*. This is particularly important given the rate of change we experience in business at large and most particularly in technology. It is important to keep two significant references under perspective:

- The lifecycle of a parallel processor today is 1½ years – a far cry from the 25–30 years of mainframes.

- A NASD (National Association of Securities Dealers) study found that to survive, a Silicon Valley firm must reinvent itself every 2½ years.

A good example of the rapid product lifecycles of modern technology is offered by Hewlett-Packard. In any one year in the 1990s, more than half of the company's orders were for products which did not exist during the preceding two years. The average lifecycle of the company's products now stands at less than 4½ years.

By contrast, in the late 1970s/early 1980s, the lifecycle averaged 7 years.

To the untrained eye, the different ratios given in the preceding paragraphs seem to divert from one another. But in reality they confirm the same trend which characterizes companies which wish to survive:

- Faster and faster product changes hitting the market.
- Greater and greater power at an affordable cost.

There are plenty of things we can do with what used to be Cray computer performance which is now offered at desk level. And we can get significant competitive advantages if we capitalize on the developing new opportunities. Let us not forget that these are available to all our competitors – therefore we should lose no time in exploiting them.

CONSUMER ELECTRONICS, COSTS AND OPPORTUNITIES

Thanks to the power of the microprocessor, during the decade of the 1980s computers became an indispensable part of the way of life in the business world.

Microprocessors also got themselves embedded in consumer electronics. Few people realize how this development changed our way of living and working.

- Fifteen years ago it was unusual to find anyone, outside the rather introvert world of data processing, who used a computer screen on a regular basis.
- Today, the majority of people involved in business management rely on a workstation to do their job.

Provided the software-development bottleneck is broken, the same pattern will repeat itself in the late 1990s with desk area networks and other parallel computers. Those organizations will be best fitted to profit from vast computer power who prepare themselves *now* for the challenge.

Preparation means first of all ability to meet the *cultural change*, of which we spoke so often. This can be done through lifelong *re-education*. Man is a re-programmable engine. But preparation also means a huge attention to *costs*.

- For years the cost of data processing was not a pressing concern for user organizations,
- But during the last few years this reference has been turned upside down.

Costs are incurred to reap benefits, and in this connection the new generation of computers and communications, as well as the broader perspectives in advanced applications, go hand-in-hand. We must always account for the fact that high-performance computers and communications systems made possible by increasingly powerful semiconductors, place new demands on applications development. In so doing, they:

- Reshape part of the corporate world that has itself been in a process of change during the last two decades, and
- Provide unexpected opportunities for the firms who know how to capitalize on an advanced implementation environment.

The able satisfaction of managerial and professional requirements propelled by the market as well as by competition, and a stringent quality specifications for products and services, are becoming a path for challengers to take away business from the current leaders.

The business opportunities of the 1990s are ideally suited for imaginative strategies in technology, manufacturing, marketing and finance. Companies able to rapidly capitalize on what technology makes possible have gained significant advantages – as the use of workstations, local area networks and client–servers documents (see also Chorafas, 1994b).

Since the early 1990s, the workstation, personal computer and microprocessor markets became one – but then they fissured again along a different axis of reference. Attention has been paid to two fronts:

- Top-end workstations, and
- Portable-computing.

A thin, flexible line divides these two domains; not a heavy line. Therefore, workstation and personal computer architects are using essentially the same techniques flowing in the same evolutionary direction. With few exceptions, the only remaining difference between the design disciplines of RISC and of X86 is their instruction-set architectures – and this too is changing.

- Implementations are fairly similar,
- Operating software and user environments are converging, and
- The instruction streams the chips execute are also becoming similar.

In other words, the new dividing line that has developed in the microprocessor industry is the distinction between desktop and palmtop computing, leading to novel design philosophies within each line which have a great deal to do with communications.

Just as significant in terms of future costs, but also business-type implementation in virtual office environments, is the merging of industrial and consumer electronics. This fusion of computer technology has made market definitions difficult, even if:

- Some companies look at themselves as computer makers,
- Others as consumer-electronics firms.

However, both are faced with the challenge of breaking new ground, and they have to move out of saturated markets into new areas. A consumer-electronics product is any one used in the household. Some define it as a product that can end up in the driveway or inside a home and the bill can show up on a personal credit card.

What about the further-out goals of HPCC? Is a US government project going to finance consumer electronics? I can see that many people in the HPCC programme would try to brush-off this query by saying that this is not the market targeted by the work they are doing. But if they think so, they are wrong.

I have already brought the reader's attention to the fact that the markets for client–servers, desk area networks and parallel computers overlap. Reference has been made to client–servers providing *supercomputing services* – at multiples of 120 MIPS or more – supported by commodity chips.

- Desk area networks will magnify this comparison, and cut squarely into the HPCC goals – at more than an order of magnitude lower cost.
- By so doing, they will undercut massively-parallel computers in large chunks of the market.
- This will tend to make the HPCC products look like being too little, too late – at too much cost.

MIPS, costs and business opportunities go together. The mainframers never understood this issue and that is why the market has been slipping out of their fingers. The disparity between a pie in the sky and what is available at an affordable cost is between the ideal and the real – which essentially boils down to the promise and the product on hand.

COMMODITY CHIPS HAVE BEEN USED WITH MIT'S J-MACHINE

What was said earlier about consumer electronics, costs and opportunities has recently been documented by a new massively-parallel computer, known as the 'J-machine'. It has been developed by MIT's Artificial Intelligence Laboratory, and it can solve problems not only faster but also less expensively than a more traditional parallel processor.

The J-machine's processor-intensive design is viable because, as we have seen, *commodity chips* in recent years have become both cheap and plentiful – permitting a great increase in *the speed-to-cost ratio for solving problems.*

- Over the years, the processor proportion of a computer's cost has been dramatically dropping.
- The consumer market aftermaths we have just discussed further sustain this already established trend.

One of the innovative features of the J-machine is its internal communications. It uses *wormhole routing* as a transmission mechanism (which has also been used in the German project by GMD-FIRST known as MANNA), which allows a node to start passing along a message to the next node before it has finished receiving it from the sending node.

The J-Machine currently incorporates 1024 nodes, each consisting of a computer processor chip and one megabyte of memory. This process of hardware partitioning is supplemented by a message-driven processor (MDP) design.

- MDPs are packed onto 16 boards with 64 MDPs apiece and stacked so that each node can communicate with its four neighbours on the same board and the two above and below it.
- Communication between the MDPs on separate boards is made possible by conductive rubber strips compressed between the boards.

Messages travel via a series of fine silicon lines embedded in the rubber, two one-thousandths of an inch apart. It takes an average of only 1.5 microseconds to send a message between nodes in the J-machine, compared to upwards of 1000 microseconds in Ethernet-linked computers.

The equipment in reference also has 16 hard-disk drives of 400 megabytes each. The MDPs have been manufactured by Intel to meet the design requirements of the MIT group, and are based on Intel's 25-MHz 486 chip, hence they are commodity microprocessors. These can be changed to Pentiums, P6 and other chips steadily coming out of R&D in a fast time-to-market flow.

Upgrading at systems level without upsetting current designs and the applications library is a brilliant strategy. It effectively makes it possible to capitalize on the growing power of the microprocessor which – for the 8088 chip to P6 – has gone up by three orders of magnitude, as Figure 9.3 demonstrates.

Graduate students designed the J-machine, which is a telling reference to what can be done by learned people and tight budgets. But there is also another reason why HPCC sponsoring agencies, executives and laboratories financed by this programme should take good notice of the wind of change.

Figure 9.3 Intel's accelerating chip speed, as measured in millions of instructions per second

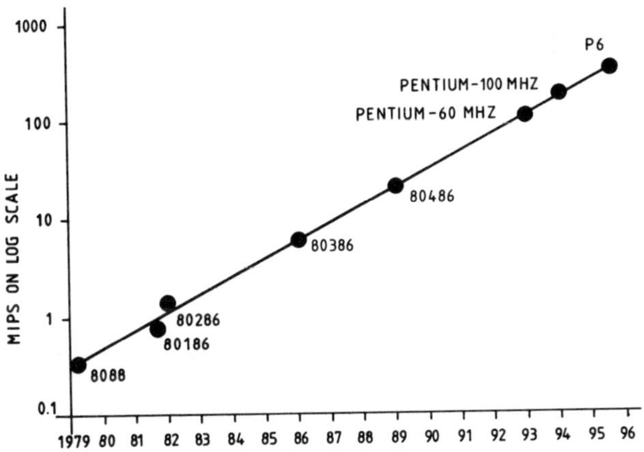

- The federally-sponsored Superconducting Supercollider project being built in Texas was struck out by Congress, even though the huge tunnel was already built (see also Chapter 11).
- The mood in Washington is no more favourable to white elephants, particularly those who lose the track they should be following, and therefore cannot deliver.

It is not enough to suggest in Symposia and other conferences that within the HPCC programme both software and hardware developments will be continuously aiming to provide higher quality. Higher quality, tighter tolerances, more reliability and greater online non-stop processing capabilities should *de facto* characterize *all* projects.

In a managerial sense, it is more sensible to worry about the fact that goals which look rather disjointed do not come to fruition and therefore represent rather wasted efforts. In this age of mass production, mass-marketing miniaturization and significant R&D budgets, focus has become the keyword – and the same is true of deliverables.

We have often spoken of the key role software plays within this broader reference. As only the leading-edge organizations are able to appreciate, a crucial part of this effort is mastery of knowledge

Figure 9.4 The continuing use of classical progressing tools increases costs and development time exponentially knowledge engineerings can bend the curve.

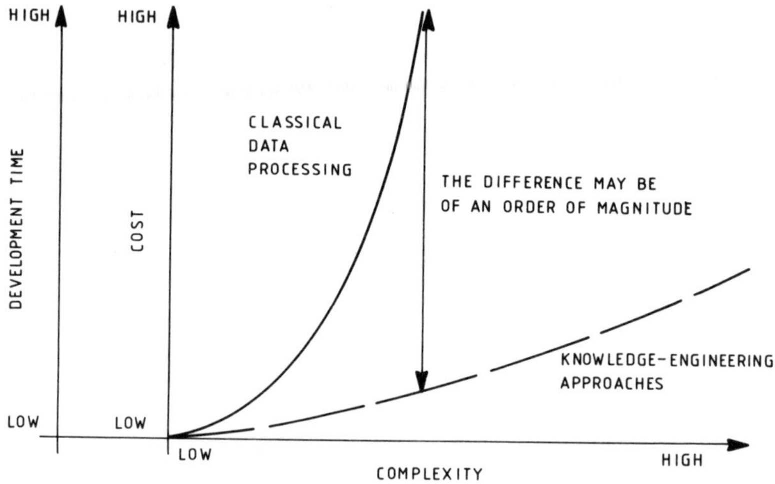

engineering to enable the wider use of experience gained in the foremost implementations which have seen the light so far.

Based on a research project I did in 1994, Figure 9.4 presents a convergent opinion among business and industrial firms on how the level of complexity characterizing current data-processing applications can be broken through knowledge artifacts. This reference is valid in regard to both *development time* and resulting *cost*.

This can be stated in conclusion. In order to be really successful, HPCC, and any project similar to it in Europe, must pay a great deal of attention to business-oriented performance criteria. The end of the cold war, tax-cutting, the budget deficit and deregulation see to it that government funding ceased being an entitlement programme.

The message is that without new departures and imaginative approaches to business-oriented applications issues, it will not be possible to effectively utilize the orders of magnitude of great computing power being developed – and therefore justify more funding. Yet, the able utilization of such power is crucial both to competitive positioning and to the project's own survival.

10 Research Projects in Telecommunications and Parallel-Computer Programming

INTRODUCTION

Programming and coding concepts, as well as tools and processes, connected to the use of computers have evolved quite significantly since the 1950s when computers were first marketed as commodities. They have also changed from notions prevailing in the early 1970s, when computers started being used in private branch exchanges (PBX) and in switching centres. Parallel-programming models have developed from procedures devised for specifying systems in which:

- There is a collection of runtime concurrent processing elements, and
- Different processes compete during operation for common resources and services.

While this evolution towards parallel programming has been in process, and clearly affects telecommunications, little attention has been paid to the fact that the prevailing view on the behaviour of a concurrent system conflicts with the one of a truly parallel solution in which processes *cooperate* rather than *compete*.

Many of the programming problems faced today in advanced switching systems and in multimedia applications, result from this wrong interpretation of *parallelism* which has its origins in multiprocessing done through mainframes and dating back to the late 1960s.

- Fine-grain parallelism has nothing to do with such obsolete multiprocessing concepts,

- But this has not yet slipped into the thinking process of major vendors and many user organizations.

Knowledgeable systems specialists debit this oversight in program structure and resource employment to the fact that a cultural change has been slow in coming. As a result, there is a conflicting view in parallel programs – on the hypothesis that it is extremely difficult to combine efficiency, portability and robustness.

Such difficulties can only be overcome if parallel-programming models are devised that allow a higher level of abstraction combined with the able handling of databases. Both issues are what the ADAMS Project at the University of Virginia aims to achieve, hence the interest devoted to this project in the present chapter. (ADAMS is an acronym of *A*dvanced *DA*ta *M*anagement *S*ystem, designed to function as an interface between user processes and one or more persistent databases.)

In a computational sense, ADAMS extends into parallel processing a paradigm which has already been used within the database community. It advocates an enriched relational model for data storage and retrieval, which features inherent object-oriented approaches – therefore combining the extensional and intentional database aspects we studied in Chapter 8.

- The project at the Institute of Parallel Computation, University of Virginia, is by no means a unique case.
- However, it does represent a good example of carefully done research on software, which contrasts to the many negatives we will review in Chapter 11.

What we will see in late sections of this chapter has a direct impact on communications, particularly in advanced research projects. This will be demonstrated in the next section which brings under perspective the goals and structure of Japan's Advanced Telecommunications Research (ATR) Laboratory, and the projects which it has undertaken.

IMPLEMENTATING PARALLEL PROCESSING SOLUTIONS TO COMMUNICATIONS PROBLEMS AT ATR

In order to position themselves against the forces of the coming decade, the Japanese founded the Advanced Telecommunications Research Laboratory. It was established on 22 March 1986, with support from industrial, academic and governmental organizations, to serve as a major centre of basic and advanced telecommunications R&D – and in 1989 it moved to its permanent quarters at Kansai Science City, in Japan's research triangle.

What distinguishes ATR from other laboratories is its research horizon. It is looking to the future over a 15-year timeframe. Some of the products it develops may not come to the market until after the year 2010 – and they will be marketed not by ATR but by one or more of the Japanese telecommunications companies sponsoring its research.

There is a public relations element to this. One of ATR's functions is to establish research connections both domestically and abroad. The main goal, however, is intensive research in four areas:

1. Communications sciences;
2. Information sciences;
3. Optical and radio communications;
4. Human sciences.

Parallel programming, in the way we will be studying it in this chapter, is of direct interest to all four of these research areas. Also, practically all four domains address the subject of multimedia and other issues crucial to the advancement of technology – such as, for instance, the development and use of agents.

The aim of ATR's Communications Sciences division is to achieve *advanced intelligent networks*, with systems solutions oriented to serving a growing array of needs – but, designed for ease of operation. To fulfill this mission, the Institute devotes itself to basic research in the following fields: full-sensory communications, non-linguistic communications, high-security networks, and the automated development of communications software – including *parallelism*.

The basic premise behind the mission of this division is that, until now, the goal of telecommunications technology was to establish realtime connectivity between distant locations in an economical manner. This goal having been more or less accomplished:

- From now on, the types and volume of information will increase.
- People will depend more and more on information-exchange technologies that they can use with confidence, and
- System solutions must be focused on these new realities – not on past notions.

Research on the automatic generation of communications software focuses on the need for speedy, accurate and continuous processing, while the telecom structure is becoming increasingly complex and extensive. Many diversified future services call not only for high performance, but also for flexibility to satisfy varying user demands.

Along this line of reasoning, the laboratory aims to research new means of expression, inventing and upgrading tools, to achieve major progress in software generation. The same division is experimenting on the automatic input of 3-D data and the effective use of visual information. This is done on the understanding that communications are expanding beyond the level based on speech, such as telephone, to include visual communication – including figures and images.

ATR's Communications Sciences Division also studies security and protection technology including pattern identification, picture identification (through deciphering codes), cypher logic, and picture processing. Parallel programming is crucial in many of these fields.

Work at the Information Sciences Division has different goals. It aims at the development of an automatically-translating telephone system. Progress depends on basic research on component technologies, speech recognition and synthesis, as well as computer-based translation.

In an effort to achieve the ultimate in telephone communication, these projects are carried out with component parts integrating into an experimental prototype of automatic speech translation. This, too, requires a great deal of parallel programming.

The goal of machine translation technology is to automatically handle human dialogue between English and Japanese – and to do so in realtime. Since spoken language differs considerably from written text, this laboratory is collecting a large body of conversational data, by simulating telephone and inter-keyboard conversation. Such data is used to study the syntactic and contextual properties of conversation.

The Optical and Radio Communications division of ATR seeks to establish the technology which can permit any-to-any communication, in any place and at any time. It conducts basic research on space and mobile media as well as devices based on artificially modulated material structures. Research objectives include:

- Optical intersatellite communications, expected to become the most important element in space communication.
- Mobile communications systems making exchanges possible in a global realtime sense.
- Materials and devices for new communication equipment which will be small, lightweight and multifunctional.

Research on basic technology for optical intersatellite communications poses parallel-processing requirements, and is undertaken on the premise that space linkages are a crucial infrastructure upon which human activity beyond the atmosphere will depend. Improving the quality of optical intersatellite communications helps to develop mainstream technology.

In the research for mobile communications, active array antennae and advanced signal processing are being developed in new frequency bands to overcome radio interference and fading problems. Other fields include materials science and bioelectronics.

For its part, ATR's Human Sciences Division addresses the basic mechanisms of perception and cognition. The human senses of sight and hearing are investigated from the engineering, psychological and physiological standpoints with the goal of obtaining a breakthrough in the technologies of visual pattern and speech recognition. Work also seeks to develop efficient and friendly man–machine interfaces.

The premise is that although telecommunications equipment has made remarkable advances, it is often difficult to use. The reason is that current machines are designed mainly from an engineering perspective, hence they are not properly adapted to human nature and intuition. The solution to the problem lies in intelligent man–machine interfaces, including the emulation of behavioural mechanisms.

The visual-pattern-perception department of this laboratory studies human visual-information processing, learning lessons to be applied to automatic pattern recognition. This issue is examined through psychological and physiological experiments, as well as by means of computer simulations of visual models – which are more effectively studied through parallelism.

As these examples help to demonstrate, there are plenty of opportunities for parallel processing in the communications field, particularly among advanced research projects. At the same time, as we will see in the following sections of this chapter, able solutions to parallelism are doable – provided those responsible for implementation have the imagination to apply technology in novel ways.

ENHANCING CONTROLLABILITY THROUGH OBJECT-ORIENTED SOFTWARE SOLUTIONS

Object-oriented solutions are very important both with computers and with communications – which is the justification for including this chapter in this book.

Whether in massively parallel computers, desk area networks, local area networks, metropolitan areas or long haul, a distributed environment makes mandatory a software-implemented sense of controllability. A basic premise is that all pieces must be able to operate perfectly as standalones, but there should also be guarantees that they will work as a system.

Object-oriented software offers many features which can see to it that a shared-nothing architecture works as an aggregate, but some approaches which lack implementation know-how have also introduced opportunities for errors. Therefore the techniques designed to test object-oriented software must be very carefully examined.

Software controllability is not an issue which has come with object-oriented approaches and parallel programming. It is present with *all* programming efforts, but it must now be more carefully examined. Etymologically, *controllability* is a term used in control engineering to mean the ability to put a system into a desired state by manipulating input variables.

- Simply because a software construct is man-made does not ensure that people are always able to control it.

The same thing, incidentally, is true of hardware as well as organization – whether it regards governments, universities, industries, banks or other business units.

- Many efforts are made in vain to control an inherently uncontrollable structure – and this happens with surprising frequency.

The theory of controllability defines the conditions under which control is possible. In its fundamentals this theory is mathematical, but it is important also to have a concept of controllability for systems that cannot be expressed in a deterministic form. In this case fuzzy engineering may be the answer.

The use of fuzzy engineering and genetic algorithms in connection to software development is so much more important as expectations on the quality of software products are rising, while constraints on cost and human resources are tightening.

- Many companies have incorporated object-oriented technology into their software development process as a means to address this problem.
- The result is that we have attained many benefits such as reduced development time and improved reuse, but also created other challenges.

Constructing a highly available, stable and robust system using object-orientation requires new techniques and tools to test the software, particularly class-level testing. This is another way of saying that a cultural change is necessary, as well as a change in methods. Those on the supply side of software have to ask themselves some tough questions: am I really adding value, or am I simply contributing to the problem?

Dr Pfaltz, of the University of Virginia, Charlottsville, VA, and his co-workers have asked themselves this question, and tuned their development in a way to provide a sound answer with ADAMS,* and not only have they accounted for controllability but also for integration requirements.

A similar statement can be made about the programmability of parallel engines. The linguistic part of ADAMS permits the expression of parallel applications, using a programming style similar to many database programming languages currently in use.

- Some experiments have been conducted in implementing ADAMS programs using logic-based and object-oriented languages.
- The focus is to develop efficient distributed algorithms for retrieving

*ADAMS = Advanced Data Management System.

information elements on the basis of selected properties and relationships.

The reason for incorporating object-oriented approaches is that the pure distribution of structured data, such as tuple structures in a relational model, tends to be awkward in terms of parallel processing. (A matrix is composed on columns and rows. The rows are tuples). It does not present a natural granularity on which a rational distribution can be based.

Relations are commonly partitioned horizontally with individual tuples assigned to different sites. If attributes are indexed for efficient query retrieval, then:

- Either local inverse attribute indexes must be replicated and maintained at all participating sites with the query broadcast to them, or
- A single index must be copied to the processing site, with user identification (UID) prefixed with site path names.

An object-oriented model is much more flexible, but objects may include numerous sub-objects as well as comprise larger super-objects. This leads to a number of other issues.

Must all components of a single object reside at the same processing site? Should an object-oriented implementation feature objects of a class managed and manipulated by a class manager? Is it necessary to bring classes and superclasses from auxiliary (disk) storage into main memory prior to instantiating them into objects?

ADAMS aims to answer these queries and also to do something about the answers. At the kernel of this effort is the realization on behalf of the developers that, in the 1990s and beyond, parallel processing is essential if a functional approach to data representation and manipulation is to be successful. A functional approach encourages parallel computation.

DOs AND DON'Ts IN CODING PARALLEL PROCESSORS

One of the pecularities of technology has been that even the experts know very little about its many facets. This is true of computing at large and most particularly of parallel computing.

A lot of problems have to be resolved before parallel computing gains real popularity in the user community. Yet, while issues concerning, for instance, efficient coding algorithms for high-performance computing are still being investigated, scientists, engineers, financial analysts and technical programmers are very much interested in the benefits parallel computation can offer.

Users and potential users of high-performance computing have in common that their processing requirements exceed the capacities of even the fastest sequential machines. People have turned to parallel processing because:

- Problems are too big,
- The time constraints too pressing, and
- The need to experiment with the effects of nondeterminism is growing.

What mature users of computer equipment really want is *performance*. While it is a fact that parallelism is the only way to continue performance enhancements, this cannot be done without the prerequisites discussed so far in this chapter.

A constraint which creates a slow transition to parallelism is that software support is designed along old concepts. The programmer who wants to parallelize existing serial code – an unadvisable practice – is primarily interested in speedup, or obtaining the same results much faster.

- Such programs are largely production code which has been tested, used and enhanced over a period of time, but for a different architecture.
- Hence there are difficulties inherent in the conversion, even if the parallel language uses some new features.

Typically, maintenance and extension of serial software has taken place over many years and was done by many programmers. There may be millions of lines of serial code in a program – and they are often poorly documented. It is better to avoid a conversion policy because it has primarily negative results.

Since reverse engineering has given no commentable results at least so far, old code is only partly known or understood by the programmers who are now responsible for parallelization.

- In transforming the program to parallel form, lots of errors can be introduced.
- High-performance computing may also expose bugs which were latent in the sequential version.
- Existing compiler technology for the automation of parallelization is limited in scope and only rarely provides adequate speed-up.

These issues are brought into perspective not to discourage the use of parallel computers, but to caution. New technology is ideally fitted for new applications, and this principle is valid all the way to the coding level.

New applications need new tools, a principle which underlines the importance of ADAMS to which this chapter is dedicated. Whether ADAMS or other platforms are chosen, primary attention must be placed on the parallelization of new, demanding applications which require well-studied techniques and new tools for software development.

One of the major challenges is to be certain that we are effectively using the resources at our disposal. Another is to assure that the program adapts to changing parameters. Productivity is appropriately supported through prototyping and experimentation, which has the added value of cutting the design and development cycle.

One of the problems with parallelism in the development of programming code is that different algorithms scale up in different ways, and there is also the phenomenon of algorithmic discontinuities, where solutions move to another algorithm. Alert software experts are aware of this fact since it is also present with serial-type programming. Parallelism simply magnifies it.

Something we have learned over 50 years of programming practice is that many different approaches may need to be covered in the spectrum of a given application. With parallelism, software performance must be sustainable across a range of problem sizes and granularities, which adds one more dimension to the stated problems.

As with every project in software development, reliable, consistent parallel performance within a transparent user environment is an obvious objective. But let us retain from this introductory discussion that some of the problems as well as the DOs and DON'Ts are also present in the sequential programming environment.

The difference is that with sequential programming, paradigms tend to be better understood, there are more libraries of coded algorithms with known properties, and the compiler and debugging support is more comprehensive. Also, in several cases performance bottlenecks may be dealt with using some available techniques.

All this speaks volumes about the care to be taken in developing efficient code for high-performance computing. We must always exercise special diligence when tackling a new environment with many unknowns and uncertainties about the results.

ARE WE *REALLY* PROGRAMMING PARALLEL MACHINES?

One of the reasons why programming parallel machines is a demanding task can be found in the fact that generating an algorithm requires the analyst/programmer to assimilate the interactions of multiple control paths. Another reason for complexity is that synchronization and communication are necessary to avoid contention and deadlock.

Up to a point, these constraints are eased by the fact that the distribution of processors and servers in both loosely and tightly coupled systems has been well studied.

- Looking at an attribute as a function encourages its implementation as a service process, which can reside anywhere in the distributed environment.
- In a programming sense, only the name of the attribute need be bound to its present server site, not individual tuple or object identifiers.

These two points are made by Dr John Pfaltz and his co-workers, who are working on a distributed configuration in which all data access requests are handled by server processes. The latter are incurring initial startup latency with the assumption made that a portion of initial latencies can be overlapped.

As these studies tend to indicate, with a parallel distribution the computational delay (response time) to execute a sequence of attribute function assignments could be comparable to a relational tuple insertion operator. A prototype comparing data representation using structural and functional storage was implemented on an iPSC/2 hypercube, with attribute functions seen as servers and distributed to various nodes. (Within this environment, storage requirements were handled by independent storage manager servers which in turn employed Intel's Concurrent File System (CFS) for lowest-level storage.)

Other parallel computing projects undertaken by universities and in industry demonstrate similar experiences but also come up with a number of problems. Some among them have to do with debugging.

For instance, once the program is executing on the parallel machine, it may perform poorly or may not function properly. Debugging is necessary and can prove to be a rather complex issue; the more so when we deal with multiple control threads.

The programmer needs to know about the runtime behaviour of his code in order to test the hypotheses connected to his search for bugs. Typically implemented through software, appropriate tools should be available to help in specifying actions to be taken on indications provided through monitoring. Monitoring can help to direct the analyst towards:

- Imperfections in the code which should attract his attention,
- Possible sources of contention in parallel data access, and
- Logic errors in conditionals, as well as their cause and aftermaths.

Expert systems can be of help in tracing and also in displaying relevant information in a form that is easily understood. As we know from programming serial machines, however, no tool-set will ever automatically debug the program.

With parallel as with serial coding, it is the analyst/programmer's job to formulate possible causes of errors and decide on possible performance problems as well as their causes. What knowledge-engineering-enriched tools can do is to make the analysis of steps of the coding process as efficient and objective as possible.

Another contribution knowledge engineering can make in terms of parallel programming relates to the fact that such programs often depend on the:

- Order of access to shared resources,
- Change in flow of the computation, and
- Gathering of status information, such as values of program variables.

With programming for high performance computing it also becomes necessary to provide architecture-type support in the form of projecting hardware and software configuration. Just as important is to assure that the monitoring system does not affect the relative timing of different control threads, assuring both flexibility and minimum intrusion.

Communications disciplines are also vital allowing the parallel machine to be accessed from remote sites, with several users working at separate nodes but data-sharing and time-sharing concurrently the same system. Support for runtime monitoring of such parallel programs is critical throughout an implementation, not only in the debugging phase of parallel-software development.

A similar statement can be made about the need for developing and maintaining a *quality database*. Information gained through the actual execution of parallel programs is invaluable in correcting and optimizing the operations of a high-performance computer.

The last statement is particularly valid in the case of a dynamically reconfigurable architectural design. For example one made for a parallel processor with nodes which may be partitioned to form independent or cooperating sub-machines of various sizes. To switch between consecutive modes at runtime calls for:

- Instruction level granularity,
- Tuned to face the requirement of parallel computation.

Online distributed database access is a 'must', which explains the particular interest placed on database management by the ADAMS project. Attention to detail is also necessary to support sophisticated logging operations with automatic occurrence-time recording added to the program's primitives.

A further goal is to gain information about the occurrence of events of interest during the execution of the parallel program. Such events are subsets of those occurring during the execution, and knowledge of their occurrence can convey particularly useful information in terms of systems control.

SOME BASIC ASSUMPTIONS BEHIND THE ADAMS RESEARCH PROJECT AT THE UNIVERSITY OF VIRGINIA

One of the premises behind the ADAMS project is that in spite of efforts at standardization, heterogeneous data systems will remain.

At the same time, as more interdisciplinary investigations are mounted, there will be greater pressure to analyze data drawn from disparate disciplines,

The dual effect of the two trends which have been mentioned leads towards the need to make multiple information elements comparable,

- Converting data sets to local site format, and
- Creating multivendor distributed database systems.

In turn, this brings into perspective the requirement of performing research directed at solving the problems of heterogeneous data-management environments with particular emphasis on their usage in parallel-processing environments.

According to this frame of reference which is applicable to all fields – from pure science to banking – today the analytic ability of a professional is impaired when an information element received from another site is in a format incompatible with the local system. Subsequent conversion of information elements (IEs), the University of Virginia project postulated, depends upon:

- Adequate metadata to interpret the format and structure of the foreign data, and
- The conventions expected by the local processing system and their timely execution.

Without metadata, hence knowledge-engineering approaches, analysis involving multiple information elements from disparate sources can be very difficult. Even within the relational world, tuples and relations obtained from Oracle, Ingres or other DBMS need not be immediately comparable. The problem is magnified with data coming from less-rigid data models.

At a much deeper level of reference, the outlined problem involves the general issue of *data fusion*, which must take into account the semantics (intended meaning) of required information elements in order to make meaningful comparisons. Hence the new approaches being studied.

'Our concern is scientific databases', said Dr John Pfaltz, 'and the relational model is not effective.' He took an example of two projects: NASA's work with wind-tunnel results and CAD/CAM programs for spacecraft written in C, which do not talk to each other – yet they are related.

To solve the problem, the Institute of Parallel Computation has developed in C++ the ADAMS Interfaces Language with type hierarchy and an object-oriented flavour. The basic premise rests on the fact that all programming languages have I/O statements which are not really part of the language's core. Hence the goal has been to:

- Give the programmer an embedded ability to interface with the shared data space, and
- Do so without constrained, closed solutions which are too restrictive.

This approach was chosen on the premise that the current heterogeneity in data and operational requirements is a fact of life. Experience so far has demonstrated that it is unreasonable to expect all databases to converge on some unifying standard; hence heterogeneity will continue to be a force to be reckoned with.

ADAMS is therefore conceived as a citation mechanism which would allow end-users to locate and examine precisely the data needed in an investigation. It was noted, for instance, that in a scientific research project (and by extension, in analytical and mathematical work in finance and other fields of implementation):

- Much of the interesting metadata is actually citations into the scientific literature.
- Such citations can be used to help assess the quality of data which is being browsed.

Many prerequisites exist to such work and the ADAMS project recognizes that a crucial part of this effort is the creation of a data dictionary with:

- Local,
- Private (user),
- Shared, and
- System information elements.

This 4-level approach helps to differentiate the System IEs, namely those everybody would use, from the Task IEs which concern a specified collection of users. The private user level provides further detail on the task reference, while the local level is non-persistent.

The research on ADAMS is both object-oriented and relational. But Dr Pfaltz was to comment that one of the biggest traps is to call everything an object. A value, for example, is different to an object.

- A class keeps its value, but
- Instantiations, hence objects, disappear when they perform their intended functions.

Hence the wisdom of making a distinction between the abstract class and various individual sets of objects. The design of ADAMS respects these premises, to which it adds its expertise from practical applications in parallel computing.

AN INTRODUCTION TO THE ADAMS LANGUAGE

The ADAMS interface language was created in response to a need not fulfilled by existing database-oriented linguistic constructs. This statement is particularly true with regard to supporting newer distributed-processing architectures, such as a hypercube.

- ADAMS provides an interface between host-application programs and a space of persistent objects in the database.
- This interface consists of ways to describe a user's space together with mechanisms to access information elements.

Typically, the accessing mechanism is served by a query language (data manipulation language, or DML). The description of the user's space is made by the data description language (DDL).

ADAMS has not been intended to be a complete language by itself. It has been designed to provide a clean interface between applications written in existing programming languages, such as C, C++, Pascal, Ada, Fortran; and to help manage shared as well as non-shared information elements.

Since ADAMS is a linguistic construct for data representation and access, it can be embedded within a host language, such as C. This makes it necessary to:

Figure 10.1 Preprocessing and compiling of ADAMS statements

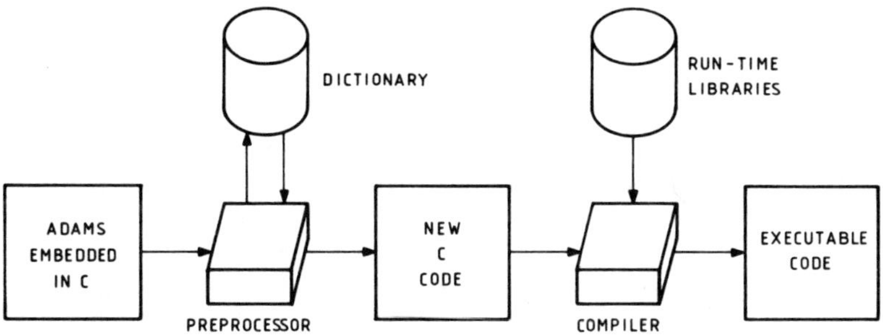

- First pass any program containing ADAMS statements through a *preprocessor* as shown in Figure 10.1.
- The preprocessor converts every ADAMS statement into a sequence of procedure calls (in host-language format) on the runtime system.
- By scanning the source program the preprocessor creates lines of source code.

As an ADAMS statement is parsed, its corresponding line is actually a parse tree passed on to the code generator. The latter converts the ADAMS statement parse tree into the corresponding sequence of host-language runtime procedure calls.

The output is a program consisting only of host-language statements. The latter incorporate the ADAMS contributions (that is, a *parallel-processing* system implementation), a functional approach to data representation, O-trees (explained later in this chapter), and a cache coherence mechanism.

As expressed by the Institute of Parallel Computation, the overriding goal in designing ADAMS has been to create a flexible database system that would actually be used by a large number of scientific application programmers:

- In ADAMS the defined meaning of a series of bits is known as a *codomain*.

One may have a codomain consisting of real numbers, or of all strings beginning, say, with the letter 'X'. Also, image spaces and attribute functions.

- Another basic concept is *class*, in an object-based sense.

Every callable information element must belong to a class which represents its structure and properties. All individual elements, or instances, within the class share the same structure and properties.

- Functions, defined on ADAMS elements, are distinguished according to their *attributes*.

An attribute is a single valued function whose domain consists of elements in one or more classes; and whose image space is a class.

Any element instance belonging to either a user-defined class, attribute or map, can be named. Such an information element has, so to speak, an independent existence. Specific values in a codomain cannot be named. They have no independent existence, but the current value of an attribute function is acting on an element instance.

In its design, ADAMS attempts to compensate for certain missing links. The relational model does not require a map concept, which is at once both its weakness and its strength. It is a strength because:

- Mapping concepts are unfamiliar to many database users (and some of the developers), and
- A database model which is based on only flat tables is much easier to explain and to visualize.

But in a real-life implementation relationships must be created between information elements, and the relational join is not always the most effective way of establishing needed relationships. This can be a major weakness.

By contrast, with ADAMS an attribute function defined on an element functionally associates a single codomain value with that element. Attributes are functions from a class of elements into codomains; maps are functions from a class of elements into another class of elements.

Not only object-oriented but also relational, hierarchical and Codasyl databases can interoperate using these ADAMS primitives. They provide mechanisms for defining and naming persistent database structures on permanent storage media; also for accessing these structures.

ADAMS uses names, or *literal identifiers*. All names are recorded in a dictionary together with their meaning. The names used to designate codomains, classes, and information elements are themselves persistent and may not be freely re-used in separate programs.

From an applications viewpoint, ADAMS is not concerned with data processing. Its focus is on the storage and retrieval of objects and sequences of information elements. The system can bootstrap itself in a manner analoguous to that of an object-oriented programming language.

Such facilities enable the database designer to name complex data constructs and treat them as basic elements in his database view. The approach is valid provided the user is endowed with rich linguistic and meta-linguistic concepts. The latter are supported through a language specifically written for programming parallism known as MENTAT.

PROGRAMMING PARALLELISM THROUGH MENTAT

As the opening sections have outlined, common to both the high-performance parallel architectures and the more loosely coupled systems is the problem of writing software to exploit available computing resources. Even for new applications, software development for parallel machines can present difficulties due to:

- The low level of abstraction at which programmers must operate in order to utilize parallel resources, and
- The inappropriateness of such abstractions for realizing parallelism in an effective overall sense.

This is particularly true in situations where parallel computers are programmed with traditional languages such as C and Fortran, extended with library calls to support shared memory or message-passing paradigms. The lack of tools for parallel coding means that the programmer is responsible for managing:

- Problem decomposition,
- Scheduling routines,
- Communications processes, and
- Synchronization.

Yet, we know that efficient exploitation of parallel resources requires high-level language support. The latter must make it feasible to manage the stated aspects.

Needed are language abstractions that permit the programmer to express algorithms by capturing domain knowledge of appropriate atoms of computation. The system solution must provide the compiler with enough information to automatically detect and manage parallelism.

Designed in response to these prerequisites, MENTAT is a linguistic construct, currently running at the University of Virginia. Its goal is to help in writing parallel software for parallel and distributed computer systems. It:

- Assures high-level abstractions for parallelism that closely match the way programmers are accustomed to writing programs, and
- Provides a machine independent, portable set of runtime support with virtual separation between the user and the physical system.

MENTAT features location transparency, object-oriented design, parallelism encapsulation, spatial and temporal uncoupling, automatic detection and management of communication, as well as synchronization. It also has embedded dynamic process creation.

The chosen approach combines a medium-grain, data-driven computation model with the object-oriented programming paradigm. It also provides automatic detection and management of data dependencies.

A graph-based, data-driven computation model supports parallelism with simple decentralized control. The use of the object-oriented paradigm permits the hiding of much of the parallel environment from the programmer. There is also a message-passing distributed orientation.

Design-wise, the MENTAT tool has two primary components. Both feature parallel-processing orientation:

- The MENTAT Programming Language (MPL), and
- The MENTAT runtime system.

MPL is object-oriented, based on C++ and able to mask the complexity of the parallel environment from the programmer. The granule of computation is the *class instance*, which consists of contained objects, their procedures and a thread of control.

MENTAT extends object encapsulation from implementation and data hiding to include encapsulation of parallelism. The latter takes two forms:

- Intraobject, and
- Interobject.

Intraobject encapsulation of parallelism means that callers of a MENTAT object-member function are unaware of whether its implementation is sequential or parallel.

Interobject encapsulation of parallelism sees to it that programmers of code fragments (object-member functions) need not concern themselves with the parallel-execution opportunities between the different object-member functions they invoke. This helps in simplifying their work.

The underlying assumption behind this approach is that the programmer can make better decisions regarding granularity and partitioning, while the compiler manages synchronization. This eases the task of writing parallel programs, making parallel architectures more accessible to the current population of computer specialists who have been trained to serial thinking.

B-TREES FOR SEARCH IN A DATABASE SYSTEM

The B-tree indexing method is somewhat similar to, but more effective than, the index sequential access method (ISAM). This discussion on B-trees is necessary to explain the role of O-trees in parallel processing, discussed in the next section.

Invented in 1970 by R. Bayer and E. McCreigh, retrieval through *B-trees* has become a *de facto* standard for the handling of single attributes from an ordered set. There is, however, the requirement that all keys must appear in the index blocks, which results in a major storage overhead.

Some studies have demonstrated that the indexes of a fully inverted file can easily exceed twice the storage of the information

Figure 10.2 Tree structures and subtrees

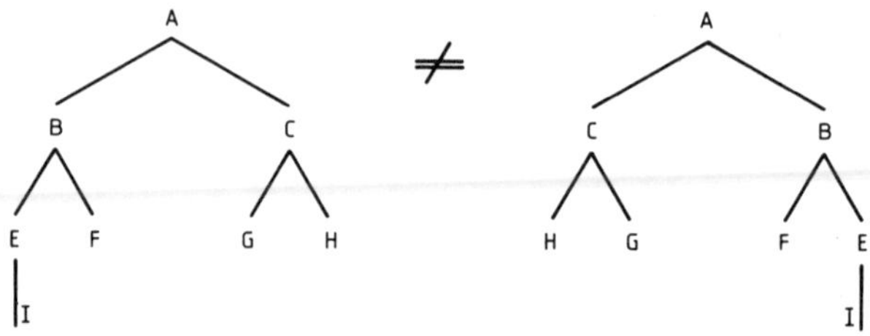

elements themselves. This creates a storage overhead and therefore a severe constraint over access performance. That's the bad news.

The good news is that mathematically, the B-tree is one of the fastest disk-based indexing techniques available. For instance, the real power of LinkIndexing lies in the efficiency of the B-tree:

- LinkIndex builds and maintains an index file that is organized as a B-tree,
- Such a file is composed of keys. Searching for a key is done by searching an index file, hence reducing the disk accesses.

In the manipulation of relatively small, centralized databases, insertion and deletion of keys within the B-tree are done nearly as efficiently as searching. A B-tree maintains its keys in sorted order, thus eliminating the time required to sort a file into order by a given key.

Over and above the forementioned advantage is the fact that the tree concept is fairly simple. As shown in Figure 10.2, such a tree consists of a branching structure between nodes. It can also be a finite index of one or more nodes distinguishing:

- One element of the set which is the *root*, from
- The remaining nodes partitioned into disjoint sets.

Each one of these sets is in turn a tree; but it is also a *subtree* of the root. One of the strengths of this approach is that it is easy to prove theorems about properties of trees. We talk of an *ordered tree* if the relative order of the subtrees B, C . . . is important. The example in Figure 10.2 is that of an ordered tree, therefore the two cases presented in the graph are not mappings of one another.

The *terminal* node (I in Figure 10.2) is a *leaf* and has degree zero. The *degree* identifies the number of subtrees of a node. A *branch* is a non-terminal node. The root is the *father* of the roots of its subtrees. Roots of subtrees of a node are *brothers* – and they are the *sons* of the root note. In Figure 10.2:

- A is an ancestor of B, C, D, E . . .
- The latter nodes are descendents of A.

A *Binary Tree* is a finite set of nodes which either is empty *or* a root. Variations on the binary search technique have been used successfully for many years, and have a place in data manipulation. The drawback is that additions and deletions to the file can only be performed in *batch type* operations because the file has to be sorted.

Searching with a B-Tree is similar to binary searching except that instead of dividing the file by two, the file is divided by some number 'n'. The greater efficiency has to do with the fact that it is the B-Tree index file which is searched rather than searching the data file. The advantages of B-Tree over ISAM is that:

- Insertion and deletion times are relatively efficient even in the worst case,
- A B-Tree index file does not need to be reorganized like an ISAM file because no size constraints are placed on the index or data file when using B-trees.
- The B-tree maintains the keys in ascending order thereby, for many applications, eliminating costly sorts.

But the B-tree solution itself has its limits which become apparent as the database size increases, and most particularly with distributed database structures and parallel processing.

Such limitations bring into perspective the interest in a compact form of B-trees which replaces all keys in its index blocks by one-byte surrogates. This index structure has been called *O-trees* for 'optimal tree' approach. (Or Orlandic-trees after the name of Ratko Orlandic who developed the concept in a graduate project at the School of Computer Science, University of Virginia.)

O-TREES BY THE ADAMS PROJECT FOR PARALLEL APPLICATIONS

In the background of the O-tree concept is the provision of a universal identification using a *one-byte index* which can be cached. O-trees represent an evolution over the standard B-tree indices and can be regarded as:

- Transformed binary trees, or
- Compressed prefix B-trees.

Their implementation helps to reduce index storage overhead by eliminating search keys from the index altogether, except for the single one-byte surrogate. This is important inasmuch as for efficiency reasons every node of a distributed database has the index for the entire data set, even though much of it will reside elsewhere. (However, such a comparison can be deceptive because the B-tree bit strings should in fact be byte strings.)

As the ADAMS project demonstrates, O-trees help in reducing the depth of the index tree. Also, since O-trees are key-length and type variant, only a single software version need be implemented in the system. Furthermore,

- The O-tree index is much smaller.
- There is one additional entry, a nil entry, in the O-tree index constituting a dummy leaf.

It could be possible that the presence of these additional dummy leaves might override any space savings obtained by key compression, but analytical proofs seem to have dispelled this possibility.

Prototyped procedures using compact representations as the access method to very large databases were implemented in C by the University of Virginia. Initial tests, involving files of 20 000 to 40 000 records, substantiated the theoretical results even over non-uniformly distributed keys.

The researchers compared O-trees with a commercially available B-tree software using real data. For a file of 10 000 actual 159 byte records with 16 byte keys, DEC's relational DBMS (RMS) generated:

- A 518 block B-tree index, compared to
- 196 blocks in the corresponding O-tree index.

This compression to 37 per cent of original size occurred even though RMS employed state-of-the-art compression algorithms and attained nearly 100 per cent utilization of the leaf blocks through B-tree optimization techniques. In fact, it is the cost of index maintenance that discourages the use of fully-indexed relations in relational databases, even more than the additional storage overhead.

As outlined by the University of Virginia, compact O-tree representations reduced overhead and led to software simplification. They also featured somewhat faster access times. Consequently, O-trees were made the kernel of functional operators in connection to the ADAMS project.

Another contribution by O-trees is the ability to cache the index. For reasons of efficiency, commercial databases are prone to cache the indexes in memory by using software implements. But when multiple processes in a distributed system are updating the same information elements concurrently, all processes must effectively be coherent.

- If two processes independently update their cached versions of the same portion of an index, the version that gets written to permanent storage must reflect both updates.
- One effective way of attaining this is through algorithms for managing locks on B-Tree nodes, but lock management in a distributed system is expensive and, at times, insecure.

By contrast, O-Tree cache coherence requires no locking, substituting it by retrying an update when coherence is violated.

As experiments have demonstrated, even with 40 processes independently updating the same index, this retry overhead is significantly less than the minimal locking required in previous approaches. Thus, an O-tree solution permits parallelism to be increased without a corresponding increase in auxiliary message traffic.

This approach may prove to be particularly suited in an adaptive non-interfering checkpointing system as required for *long transactions*. Long transactions typically address distributed heterogeneous databases and call for checkpoints to be performed in order to save a consistent state of the information elements on a separate secure device.

In case of failure, the stored data can be used to restore the database. It is however desirable that transactions are executed in the system concurrently with the checkpointing process. Hence the interest in using non-interfering checkpointing algorithms with adaptive techniques.

By supporting multiversion distributed information elements, the solution which we have seen in this section is of service with parallel computers. By the same token, it offers a valid way of approaching the requirements characterizing multimedia system – albeit not an exclusive one.

Performance of distributed multimedia databases is enhanced through synchronization using multiple versions of data objects, in addition to efficient replication. By providing a succession of views of information elements, we reduce:

- The conflict probability among transactions, and
- The possibility of transaction aborts.

This can be said in conclusion. Efficient solutions to parallel-programming challenges are not going to happen overnight through one great leap forward. Rather, they will require that a number of building blocks in a variety of computers and communications projects are being put in place. An example has been given from advanced research projects in the communications industry.

11 What's the Return-on-Investment on Taxpayers' Money Spent on HPCC?

INTRODUCTION

An article in the Science and Technology section of Business Week (1993) noted that R&D managers are having to focus research more clearly on the step beyond invention, by commercializing discoveries. 'I want to still do fundamental research, but I also want people to understand that they are doing it for a purpose', says Nobel laureate Arno A. Penzias, vice president for research at AT&T Bell Laboratories.

The last reference identifies both the strength and the weaknesses of massively-parallel computing programmes the way they are put into practice in America, Europe and Japan. As already noted, there are two fundamental areas of applications for HPCC:

1. The one involves the more-evident problems in science and technology, which are better appreciated by the parallel-computer researchers.
2. The other addresses the issues that basic researchers appreciate less, but which are even more important in successful marketing. Namely, management and logistics.

The take-off problem is that while grand challenges exist both in category 1 and in category 2, it is the business opportunities in class 2 which have the potential to create a *mass market.* But university research projects sponsored by the government, as well as the parallel-computer vendors, are not positioning themselves to face the requirements of user organizations in the broader market sense.

As we will see in this chapter, there is still another problem. In America, Europe and Japan government funding also defaults in terms of priorities. The failure to give top priority to sophisticated software at large, and new languages in particular, minimizes the results which might be obtained on the hardware front.

In terms of software breakthroughs, research, development and implementation (RD&I) are indivisible processes. They form a continuum which needs clear objectives and, as we will see in this chapter, tough project management. The most common characteristic among failed projects is that, over the years, the targets got lost from sight. There are many reasons why the directions the programmes should have taken were distorted.

BEING CRITICAL AND CONSTRUCTIVE IS AN ACT OF FRIENDSHIP – NOT A HOSTILE MOVE

In 1991, the High-Performance Communications and Computing (HPCC) Act was passed by the US Congress and signed into law by then President Bush. Subsequently, the election to the vice-presidency of then Senator Al Gore, a principal sponsor of the bill, gave the American research community reasons to feel that future funding was a self-evident truth. This, however, was more of a political aspiration than one founded on facts and figures.

Overreliance on government spending without clear return-on-investment goals and precise criteria of success, proved to be the wrong policy. In mid-1993, a couple of years down the line from the start of the project, two reports from congressional agencies analyzed HPCC research in computer architectures and reminded both:

- Researchers and
- Funding agencies,

that the sponsor's (read: the government's) goodwill and patience have limits. To continue getting funds, these audits said, research programmes must be *well-managed* and *justified* in terms of *expected results*.

Many in the academic research sector have been upset by this development. Their protests take different forms which, however, have a common background. It is described by the reaction reviewers in academe reserved to my own evaluation of where the HPCC effort was going, or not going:

- Don't rock the boat,
- Let the government's money come in.

This was simply misreading the advice I had given, and the reasons which motivated it. In the early draft on HPCC goals and non-achievements, prepared under contract with Computer Science Press (of the Institute of Electrical and Electronic Engineers, IEEE), I did not say that HPCC should not be funded, but only said that:

- It is not funded in the right places, and
- No imaginative projects around the software challenge have been undertaken.

This is more or less what the US Congressional investigation committees subsequently stated, though they also focused on another important domain on which I was lacking inside information: *spoilage and misusage*.

At the top of the list of the items brought under perspective through Congressional investigation is the fact that the HPCC contracts given so far to university professors break the implicit accord with their academic employers that:

- They will be paid relatively low professional salaries and teach students as well,
- In exchange for the freedom to work at whatever research they can raise funds to support.

Next to the list of the objections being raised by the congressional committee is that by majority academe has a *distaste* for practical problems. However, it should be noted that, in my opinion, some of the projects which do not follow this rather general rule become some of the best possible sources of radical innovation – all the way to the kind on which new industries are founded.

The secret of successful productization lies in the creation of new markets by focusing on business needs and consumer needs, rather than spreading thin over wide areas. All participants in HPCC and similar projects should appreciate that information technology is an integral part of a strategy for competitiveness – hence for growth and survival.

- Research projects must interrelate the key components of a successful strategy.
- They must be contributing to effective solutions of grand-challenge problems.

One of the basic premises underlying high technology is that *rapid progress* towards the solution of complex issues relating to users, products, customers and markets, is essential to fulfilling many of the missions of this decade. Project funding should be conditioned by the results of audits on how well basic policies and principles are observed.

This approach will be even more crucial at the beginning of the 21st Century which is, after all, only a few years away. The terms 'academic' and 'theoretical' should not be confused. Instead, emphasis should be placed on *value*. But:

- How can we assess new communications, computers and software products without describing the numerous task environments?
- How can we be sure of product success without market feedback of product characteristics such as functionality, quality and cost?

These are practical queries which help in guiding the hand of project auditors. Above all, whether corporate or government-financed, technology projects should be the subject of regular and rigorous *design reviews*, aimed to ascertain whether they are on track, spend money within budget, move fast, and produce results of outstanding quality. This is the message conveyed by Figure 11.1.

THE HPCC AUDITS BY THE GENERAL ACCOUNTING OFFICE AND THE CONGRESSIONAL BUDGET OFFICE

In May 1993, the US General Accounting Office (GAO) released a report on the Advanced Research Projects Agency's (ARPA) management of its architectural research on computers and communications. In June 1993, the Congressional Budget Office (CBO) published an analysis of the likely impacts of HPCC on the computer business and on industry at large.

Both studies on the progress of HPCC expose and exploit certain of its scientific and political weaknesses. They indicate that the pro-

Figure 11.1 Design reviews should be rigorous through the line of the project and focus on four objectives

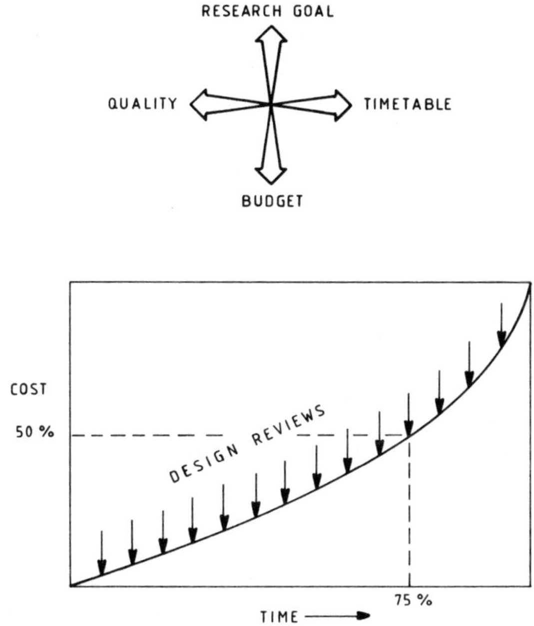

gramme as a whole, and the Advanced Research Project Agency of the Department of Defense which manages it, to a large degree:

- Overemphasize hardware and do not pay enough attention to system software.
- Have placed too narrow a range of computers at the research sites, and
- Provide very limited supervision, which hinders rather than supports the progress of the programme.

The critique is practically identical to the one this manuscript supported in its original version, to which the American research community, and most particularly the University centres, objected so much. But it is one thing to brush away the advice of peers on the need to restructure and reorient HPCC – and another to deal with the growing opposition in Congress to further funding.

The principle underlying both the CBO and GAO audits is in my opinion simple: projects which do not fulfill their mission should be killed. Governments, business and industry have to be convinced that research on advanced computing and communications technologies can accelerate the development of their products and their market appeal.

An effective use of parallel computers must demonstrate that it enables companies to integrate new knowledge and methodologies into their product and market strategies. This has to be shown in *practical* not theoretical terms:

- High-performance computing and communications technology is knowledge- and innovation-intensive.
- New developments engage the entire nation (or organization), and therefore the whole nation (or organization) should benefit from them.

These are issues that make sense, but they are generally appreciated by people with business experience – which means with the salt of the earth – not by theoreticians. By contrast, theoreticians are in for bad surprises.

After accepting the reports referred to in mid-1993, the House Committee on Armed Services asked the GAO to undertake a new, broader study of the management of HPCC research. But the committee also (for the second time), cut ARPA's research funds and directed the Secretary of Defense to report on what steps had been taken to correct the problems which had been identified.

For its part, the National Science Foundation, the second largest sponsor of HPCC after ARPA, also moved to tighten up HPCC programme planning. It formed a Blue Ribbon panel with the assignment to examine the role played by the supercomputer centres it financed, and other programme activities within the high-performance computer domain.

Where does all this leave the HPCC programme in America? As the recipients of government funds should have known and understood in the first place, it is not just a matter of spending money with a low return.

- Since this research programme is supposed to shape the future of American industry through high-performance computing and novel communications systems, it will affect the fortunes of many companies in most industrial sectors.
- Companies in all walks of life, which depend on high-techology for their survival, gain the attention of the political system – which then asks whether the direction of the R&D effort is well founded.

One of the key issues in the 1990s is that the science agencies increasingly tie support for research projects to the ability of reaching specific goals. Hence, programmes are exposed to much greater political scrutiny in terms of *return-on-investment.*

- Are the goals being funded worthwhile in terms of national competitiveness? In a return-on-investment sense?
- How well are the various programme elements tied to those goals and how is their contribution assured?
- How should progress towards the goals be measured, and how is corrective action taken?
- How will the programmes be managed to assure that focus is placed on main issues, and is not being side-tracked?

An able answer to these queries evidently means tight reins on HPCC research and development. This represents a sea of change from the policies which used to characterize military-sponsored research, particularly during the Strategic Defense Initiative (SDI) years. Hence, a new R&D culture is urgently needed.

WHAT IS THE TAXPAYER GETTING FOR HIS BILLIONS?

According to estimates made by cognizant executives and technologists in America, even with $1.9 billion pumped into the High-Performance Communications and Computing (HPCC) initiative, the US will still have a hard time catching up with Japan. One of the reasons for this assessment is that in Japan supercomputing is already used extensively in engineering design and in manufacturing.

The signs that something is wrong with high-performance computing in the US do not date from last night. As far back as August 1991 a report from the Gartner Group pointed out that Cray Research, then the principal US supercomputer vendor, had an annual revenue of less than $1 billion. In contrast, the three largest Japanese computer companies – Fujitsu, Hitachi and NEC – had annual revenues ranging from $17 billion to $45 billion.

At about the same time, while being the largest computer company in the world, IBM was still practically out of the HPCC race. In fact, in August 1991 it negotiated to purchase NEC supercomputers for resale (!). This happened while:

- IBM has had three supercomputer projects in its labs at Yorktown Heights since 1986, but none of them produced a marketable computer, and
- The junior league supercomputer vendors: Thinking Machines, Intel Scientific, nCube, MasPar and others have not been able to graduate to the major league and gain worldwide importance.

It is bad enough for all concerned, including HPCC, IBM and the junior league, that four years down the line nothing has changed. No matter if at long last it introduced a coarse-grain parallel computer, IBM is still a mainframer trying to get a place in the PC market by playing the not-so-convincing game of the PowerPC.

Neither is it to the credit of HPCC the fact that it did not give a helping hand by means of sponsoring sophisticated software, and also by financing imaginative implementation projects in business-oriented parallel computers – which will provide the applications references they need to have in order to prosper.

To a large extent this is the different vendors' own failure. They focused too much on hardware power and too little on novel software. They all hang on Fortran 90 as *the* language and missed the large business market that might have been.

Business-wise it is tragic, but it is also a fact, that practically all US computer vendors are still living in the good old days of the cold war when the taxpayer paid for the extravaganzas and hardware was king. Today, although government support is still essential, industry must get more involved in financial, product development and marketing terms.

There should be a radical departure in top-management thinking and a change at the vertex of the different supercomputer companies, who have not been able to adjust to the new realities of the market. (nCube may be the only exception to this reference because of its imaginative project in multimedia network servers. See also Chapter 6.)

- This is not just an issue of optimization.
- It hits the bottomline, and therefore it is a matter of survival.

The audits done by the General Accounting Office and the Congressional Budget Office did not get into these issues, but the critique of how HPCC is managed has been strong enough and the research community loudly objected to the two reports, just like it objected to my findings on the way HPCC has been going.

- A valid point made by the American scientific community is that the supporting government agencies seriously underestimated the importance of parallelism as a fundamental paradigm in the longer-term future.

This in itself answers, to a certain extent, the query about *where* the taxpayer's money is going and *why*. But it does not respond to the next most important query in auditing: *what* the taxpayer is getting for his money?

- The blunder that this same US scientific community made has been to focus on hardware and leave parallel software and new linguistic metalayers as a subservient issue. Actually the opposite is true.

Somehow lost in the grab for funding money are the most basic issues of any programme: *what are the real goals*? How do those goals relate to the much broader national industrial-competitiveness objectives? To the Information Infrastructure agenda of the United States?

The answers to these critical questions cannot be given single-handedly by the research community. They have to be elaborated by business and industry at large, as well as by Congress and the Administration. The best approach is a *balanced responsibility* involving academe, government, business and industry.

- There should be appreciation of the fact that this is a new territory for all of these players, and

- There exist lessons to be learned from what happened to the Superconducting Supercollider (SSC).

As early as 1992, the House voted 232 to 181 to kill the SSC, but it was saved in a Senate–House conference. However, after the October 1993 House vote, Senate supporters knew they could not rescue the SSC this time.

As a result of this abandonment, about all the taxpayers have for their $2 billion in SSC investments is a number of buildings and 15 miles (24 km) of tunnel under the Texas prairie. In reality, the loss is much bigger, because the cancellation of SSC has hit hard research in basic science.

The abandonment of SSC, along with a near cancellation of Freedom, NASA's space station, earlier in 1993, pose questions about whether the US will ever again tackle so massive a scientific enterprise. As Erich Bloch, former head of the National Science Foundation was to comment: 'There is no single country, including ours, that can afford such a big project.'

FUNDING THE BASIC RESEARCH EFFORT

HPCC, SSC, Freedom and similar megaprojects are multipurpose in their nature. Therefore, *if* conducted properly with all due care in terms of goals, budgets, timetables and quality they can have some significant and lasting results.

Funding basic research is a complex issue which has its proponents and its foes. Since the end of the Cold War, the governments in the First World progressively abandoned their commitment to basic scientific research. This has been curiously taken as being part of the peace dividend and the commitment has been made that taxpayers will only support science that produces tangible, commercial results.

- It looks *as if* during the last few years the pursuit of knowledge for knowledge's sake has fallen from grace.
- The consequence is that almost everywhere new policies are marked by the outrage of pure scientists.

Let us take three countries as an example. In America, support of science has always been more diverse in origin and therefore more robust. But 1994 seems to be the first year, for over three decades, where funding priorities have been given to research problems relating to practical solutions.

The concept is that public spending on science can and should be managed so that the tangible benefits to industry are more immediate than under present arrangements – and that they are visible. Yet, in America at least, links between academic researchers and industry are already close (some would say too close) – while there is a risk that managed research might be backward-looking.

In the United Kingdom, the government has produced a policy document embodying fairly radical proposals for the reorganization of British science. After three decades in which government support for:

- Basic research, and
- The discovery of new knowledge

was an objective of public policy, almost all the British government's increased spending on civil research is in the future to be harnessed to wealth creation. Here again, practical results dominate.

In Germany, since 1993 federal research funds are being diverted eastwards, to build up new research institutes in the new Länder (states). But Länder governments in western Germany are not able to make up the shortfall that hits their universities and laboratories. Therefore, basic research falters.

- Research funds are further reduced to finance the silly project of moving the capital from Bonn to Berlin.
- This even hurts the budget for higher education, thus robbing German industry of its future.

The Mittelstand, or medium-size companies which are the backbone of German industry, depend heavily on the universities and technical universities for research results. Cutting the budget is not the move which responds to their requirements.

Superficially, it might seem that what I said in earlier sections and what I write in this section, are two concepts which somehow contradict one another. In reality, however, this is not at all true.

- Some 10 to 15 per cent of the total budget dedicated to R&D should go to basic research. SSC is part of this domain.

Table 11.1 Goals and budget of basic research, applied research and development

Basic Research	
Goal:	Scientific discovery
Budget:	10–15% of total R&D
Hit Ratio:	5–10% successes About 15% also-running 75–85% failures
Applied Research and Development	
Goal:	Technological Leap and Productization
Budget:	85–90% of total R&D
Hit Ratio:	75– 85% successes
Particular Characteristic:	Tough design reviews and accountability

Few basic research projects succeed. In the long run the number is typically between 5 and 10 per cent. Another 15 per cent or so just survive. The balance, between 75 and 80 per cent goes to the wastebasket of time and effort or becomes part of the dust of history.

However, those basic research projects that succeed create the wealth of the next generation. They underpin new products, power new industries and generate new markets, as Table 11.1 suggests. They also provide the raw material on which will be based applied research and development projects.

- The other 85 to 90 per cent of the total R&D budget goes to applied research and development. HPCC is part of this allocation of financial resources.

The aims of applied research and development are totally different than those of basic research. Here, the goals must be concrete, the timetables stringent and the management as well as the researchers should be fully accountable for quality.

That's why the practical results to be obtained with HPCC should always be kept under perspective. There is no excuse of slippages and no room for egos. Applied research and development must be focused and should address those issues where the greatest opportunity lies. That is exactly what the HPCC programme does not seem able to do.

WHY THE US GOVERNMENT HAS BEEN BEHIND HIGH-PERFORMANCE COMPUTING?

Released in August 1991, a White House study suggested that the US can regain its former standing in the information technology market – if it tries. But it also underlined that it would be possible to do so only through the strategy of building computers based on *biological processes* and by emphasizing *artificial intelligence* (AI) systems.

According to the then Presidential Science Advisor Dr Allan Bromley, the 1990s have been federally designated as the 'decade of the brain' – and the prediction was made that applications in this domain are as promising as the first space launch. Subsequently, the Office of Science and Technology Policy reported that brain research can influence progress in:

- Massively-parallel processing,
- Cognitive AI aggregates, and
- Complex imaging systems.

All these areas underpin the so-called *Grand Challenges* of the 1990s and beyond, and therefore have been receiving significant government funding. The change of Administration had no negative impact on this policy.

The then senator Albert Gore, Jr., currently vice president of the United States, was to remark in an article (*Communications of the ACM*, 1991): 'During the 13 years I have served in Congress, I have seen a phenomenal increase in the power of computing and networking . . . Today, top-of-the-line PCs are as fast and powerful as the faster supercomputers available in the 1970s.' And the inference was made that the Government should put in seed money for more advanced developments.

Albert Gore went further to say that high-performance computers are *empowering* people. They are also empowering companies – since organizations are made of people – giving them brand new ways to process information:

- Sorting it,
- Storing it,
- Analyzing it, and
- Displaying it.

Figure 11.2 Solutions for the 1990s and beyond will rest on four pillars which among them constitute a layered architectural solution

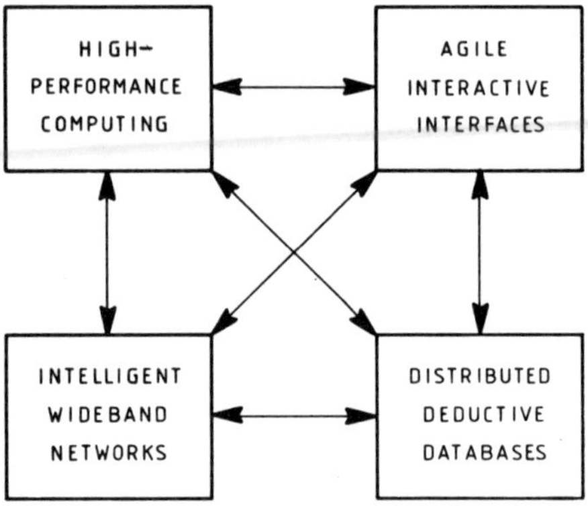

Correctly, Gore emphasized that networked computers are reshaping the world every day, from scientific laboratories to the supermarket check-out line, doctors' offices, banking services, and our homes.

Computers however, even the fastest and most cost-effective, are only one of the four pillars of the information economy of the 1990s and beyond. As shown in Figure 11.2, the other three pillars are agile, interactive human interfaces; distributed deductive databases; and intelligent wideband networks. The whole domain of computers, communications and software has to be seen as one interactive system.

Through the bill which he presented to the US Congress, the High-Performance Computing Act, Albert Gore Jr. wanted the nation to build a whole new aggregate of information superhighways made of glass fibres. The background concept envisioned the network as the first step in creating *a vital infrastructure* for the emerging information economy.

The High-Performance Computing Act funds research in how to manage and use superfast computers and networks. With its projected capacity to move multimedia information instantly across the country at low cost, the network will also serve in making the US economy more competitive – as we have seen in detail in Chapter 9.

But selling the outcome of the HPCC programme is not the government's business, even if some of the federal institutions and laboratories also act as clients apart from contributing research money and handouts in HPCC.

- Though government will remain as a client of high-performance computers,
- With military procurement de-emphasized, its purchases do not constitute a broad market base.

This brings us to the moment of truth. Whether the HPCC sponsors and researchers like it or not, massively parallel computers will be widely accepted *only when* end-users and third-party software vendors are attracted to this business. And these two populations are interdependent.

Applications-software vendors will be happy to invest in programming products for high-performance engines only when they know that their efforts can find a massive market. To sell their wares they have to make sure there is an audience of companies which are convinced that parallelism is offering them advantages. 'This is exactly the opposite of what HPCC researchers are thinking', one of the reviewers was to suggest.

SHAKEUPS THAT ARE LONG OVERDUE

Failures can happen for many reasons, however one reason underlying all of them is *complacency*. The B-2 bomber has often been criticized because of its $865 million-per-copy price tag – but most importantly it came under repeated criticism on technical grounds.

As revealed by Defense officials in public hearings, a prototype of the aircraft had failed a test of its stealth, radar-evading features. That report followed the finding of cracks in the aft deck of another B-2 prototype.

At about the same time, the US Air Force disclosed that a major new missile being developed for the B-2 bomber and other warplanes did not work properly. In talking about defects in the classified Standoff Attack Missile, one should not fail to take notice that it represented a $15.1 billion programme.

How is this money comparing to the investments made in HPCC?

In 1990, when the High-Performance Computing Act was discussed, the Cold War was not yet quite over and HPCC got a relatively small share of government money.

- The Standoff Attack Missile programme alone accounted for 750 per cent more money than budgeted for the High-Performance Computing Act.
- Yet the latter is, and it always has been, a couple of orders of magnitude more important to American competitiveness than the missile.

Every nation, and most particularly every western nation today faces this challenge. It is no more 'guns over butter' but 'high-technology over guns', and guns have ceased being the generator of know-how for leadership in the civilian markets – as was the case 40 or 30 years ago. There is a bifurcation in the two paths.

At the same time, the fact that the money spent on a government-sponsored programme is much less than that thrown at a different, lesser-value project, is no reason for losing contact with the market and with reality. If nothing else, the US can learn a lesson from IBM.

As Paul Carroll (1993) writes: IBM got to be 'like a music-publishing company run by deaf people'. The same is true of all other mainframers and, quite unfortunately of some government-funded supercomputer projects as well.

- What we can learn from IBM's failure to keep its leadership is embedded into two words: arrogance and greed.
- Interestingly enough, this is not the case of only one vendor – but rather it happens with all mainframers.

In the 1960s and 1970s, the years of the fat profits, arrogance, greed and incredible incompetence characterized both corporate and personal levels of computer vendors. Besides this, the main vendors, which in those years were the mainframers, became culturally gridlocked, thick-headed and increasingly lost in a fast-changing marketplace.

The case of IBM in the mid-1980s to early 1990s is an example of mismanagement and serious policy errors. Many companies fell out of touch with their market, but IBM is better known, hence more visible. Convinced of the infallibility of Big Blue's 'finely-tuned' management system, its executives fell out of touch with the market as the action shifted:

- From high-profit mainframes,
- To commodity-like desktop computers and local area networks.

Time and again IBM stumbled in pricing, as well as in product design. It also misguided itself through its insistance that whatever it was it offered in computers *had to be* appreciated by the market. The customers were supposed to have no other option – or opinion.

In spite of one setback after another and a market rejection which became more and more clear, IBM's mainframe establishment never seemed to comprehend the extent to which their company's clout had been fading. In the 1970s they lost the mini and maxi market and let DEC leapfrog them. In the 1980s, the same thing happened with PCs and workstations, where Apple Computers, Sun Microsystems, Compaq and others got hold of the market.

This insensitivity to client drives and the resulting IBM troubles raise questions about whether big companies can work anymore. Here is the case of an organization which used to be successful, and its downfall provides lessons for other companies, and for nations as well.

DON'T LET INERTIA RUN YOU DOWN

Big-company management and big government have much in common, *inertia* is one of their common grounds. Nobody would take the risk of exposing their career if things turned sour. Yet without risks there are no profits. American HPCC planners can learn more than one useful lesson from ICOT in Japan and the uncertain trumpet of technology of the European Union.

To get rid of arrogance, greed, complacency and inertia we have to shake things up. An organization decided on survival must be able to do what Sam Walton advised his associates and assistants: 'Be able to turn on a dime.' There is no room for worker featherbedding – much less for featherbedding in executive suites.

Walton's advice applies hand-in-glove to lean companies accustomed to shopping for value, not for high costs and buzz words.

The best way to make a success out of HPCC and the Information Superhighway is to follow Ross Perot's advice: 'Get rid of the 14th floor' – the plush executive offices at the top of General Motors' Detroit headquarters. A lean and mean national information superhighway should:

- Seed private initiative in developing increasingly more cost-effective solutions.
- Speed all sorts of products from simple to very complex,
- Give priority to sophisticated software over hardware, and
- Serve up commercial, financial, entertainment and other information services.

Not all goals will be reached at the same time, and failures can happen with any project and in any country. Only those who are totally inactive have no failures in their lives – except the greatest failure of them all: doing nothing. But some failures are not totally excusable because they could have been prevented – while in other cases:

- Projects which crash were badly planned since the beginning, or
- They have been characterized by lousy management and by big egos.

Part and parcel of a sound strategy for the 1990s is the appreciation that even if today we are a long way from where we were even ten years ago, we have only begun to exploit the technological promise of computing and networking.

- Agile software for high-performance computers and faster networks will revolutionize the way research is done in industrial laboratories and academe.
- Researchers will be able to solve previously unsolvable scientific, engineering and business problems – hence the title 'grand challenges'.

A whole new field is in the process of being created to apply the power of high-technology in aeronautics, meteorology, astrophysics, biochemistry, geophysics, economics, finance, manufacturing, merchandizing health services and many other fields.

- This effort must be supported through strategic enabling technologies such as semiconductors, superconductors and advanced materials.
- But still the greatest need is for more powerful linguistic constructs and other advanced software.

Augmented with knowledge-enriched programming products, high-performance computers and communications are today a better shield to national security than missiles and bombers. Can we learn from the failures of other efforts in order not to repeat the same errors all over again?

LEARNING FROM ICOT's FAILURE TO DEFINE THE NEW COMPUTER ARCHITECTURE

In this chapter we have seen examples from many classes of failures. Now we will take a look at a half-failure. More precisely, a project which reached one of its three goals, got itself into the wrong direction in terms of the second of its aims, and totally lost sight of the third objective it had set for itself.

Japan's New Generation Computer Technology (ICOT) project started in 1982 and ended in 1992, lasting altogether the planned years plus a couple more of additional life support. Its original goals were the following:

1. Establish a new computer technology for the 1990s,
2. Emphasize knowledge-handling, and
3. Promote parallel processing

The attainment of these goals was financed by the Ministry of International Trade and Industry (MITI) to the tune of $500 million, but the results do not necessarily correspond to costs which were incurred. In other words, the Japanese taxpayer got little return for his money.

Goals 1, 2 and 3 had been projected to support one another. In the background of the new computer technology has been an overall concept with logic programming in the kernel. This sounds reasonable, but it created a bias in the ICOT efforts because:

- A great deal of emphasis was placed on logic programming, and

- Not enough attention was paid to software needed for programming parallelism.

Under a biased perspective, developments centred around advanced versions of Prolog (see also Chorafas and Steinmann, 1989), at a time when logic languages like Lisp and Prolog had already started to move out of centre stage as the linguistic constructs underpinning knowledge engineering.

Yet the wider implementation of artificial intelligence was the number 2 ICOT goal. There is no doubt that by putting knowledge engineering to the forefront, ICOT management has contributed a great deal in promoting AI in Japan. If nothing else, its work:

- Created a two-layer concept of intentional databases (IDB) and extensional databases (EDB) (see Chorafas, 1994a), and
- Demonstrated the feasibility of deductive databases, therefore of database intelligence applied in an expanding environment.

This effort underlined the importance of concurrent logic programming enriched with knowledge representation (Project Quixote). It also framed the process of abduction and of thinking by analogy, and did some basic research on non-monotonic reasoning.

These are no minor feats yet, altogether, the new generation computer project fell short of delivering the promised results. Secrecy, among other reasons, put this project on tracks which pointed in the wrong direction. Secrecy is a very bad policy because it leads to blind alleys from which it is very difficult to escape.

Nowhere is this failure more evident than in terms of the number 3 goal: parallel processing. In spite of all the money it consumed, ICOT neither developed nor adopted innovative designs. Instead,

- It settled for an already known mesh architecture,
- Rather than a pioneering fine-grain hypercube or other new approaches.

In terms of cost-effectiveness, the Personal Sequential Inference (PSI) engine – the result of the first four years of ICOT's efforts – compared badly to the powerful workstations available in 1986 such as Sun. And ICOT had nothing to propose against the two orders of magnitude of gain in cost-effectiveness shown by the Connection Machine, Intel's iPSC and eventually MasPar.

Half-heartedly, after 1986 ICOT's efforts were refocused on the Parallel Inference Machine (PIM). But by then fine-grain parallelism was already an operating reality. Coming late and being demoralized by having missed the main goal is no way of gaining leadership. Rather, it is a sure way for accumulating losses and for falling further behind.

WHAT THE FALL OF ICOT AND OF IBM HAVE IN COMMON

In a way, the failure of ICOT to produce results in terms of architectural designs and systems able to survive and prosper, resembles what has happened with IBM – and the reasons why it fell from the pinnacle of power.

- There has been aloofness at top-management level coupled with the inability to conceive where computers, communications and software are heading.
- Gains were to be obtained without pain, once the budget was in place. In the one case the taxpayer, in the other the customers and stockholders were expected to be happy no matter how poor were the results.

Could the outcome have been different *if* it had been possible to apply retroactively the knowledge gained from this experience? The answer is that the outcome might have been different *if* the strategic plan had been originally better focused – then steadily refocused. As a suggestion, a better-focused policy might have been possible *if*:

- New types of software – through knowledge processing – had been the epicenter of the effort,
- The highly-parallel architecture had been from the start the high point, and with it cost-effectiveness.
- Emphasis had been placed on intelligent software – from computers to networks – and on the interconnection of deductive databases.

In ICOT's case, its management as well as the Ministry of International Trade and Industry, which provided the budget, was not sufficiently open and willing to communicate with the outside world. Therefore, refocusing was not really feasible.

Maybe the Japanese learned from this experience, hence the open forum which preceded the establishment of the Real World Computing (RWC) programme. In the March 1991 meeting in Tokyo, participating researchers from Japan, America and Europe set up three subcommittees coordinated by a research committee (chaired by Professor Takemochi Ishi), for:

- Basic research,
- Computer science, and
- Social impact.

Each subcommittee was organized in working groups. That of Basic Research included the areas of *brain studies*, neurophysiology, neural networks, perception, pattern recognition, brain integration, memorization, recall, genetic inheritance, learning and intuition.

The topics covered by the Computer Science subcommittee have been: optical computing and all types of networking and databasing; also parallel processing and fine-grain superparallel computing at the level of *1 million processors* that can attack a problem at the same time. (10^6 processors on a chip should not be confused with an equal number of transistors. For the year 2000 10^8 transistors on *one* chip are forecasted).

- But there is no evidence that that productization is or will be a focal part of RWC. In fact, so far RWC has failed to produce any deliverables – and I doubt it will.
- Neither is there a policy of facing the challenges posed by new departures in software – which should have had priority over other RWC goals.

There should also have been a clear commitment to bring forward what ICOT left half-baked, yet worth pursuing. For instance, the conceptual approach to *deductive databases* (which can be seen as a major ICOT contribution) and *concurrent logic programming* (CLP). What CLP still misses is a human window which will interface to the end-user by means of *visual programming*.

The goal of CLP is the development of a programming language suitable for concurrent execution. Concurrency control is provided through the Guarded Horn Clauses, with the primitives permitting implementation of parallel deductive databases through three strong points:

1. *A language construct* advancing rules of synchronization, commitment operators and use of assignment properties.
2. *Knowledge-engineering programming* featuring rapid prototyping, declarative clauses, easy debugging – and after compilation a rather efficient performance.
3. *Inheritance and meta concepts* making feasible the handling of *ad hoc* queries, mapping of high-level declarations, as well as approaches to parallelism.

Still another unfinished software effort from ICOT worth further exploiting with a view to productization is *Quixote*. This is a knowledgebank-level language particularly tuned to representation and based on constrained logic programming.

In other words it was not that ICOT was totally void of results. What has been very negative is that lip service was paid to deliverables – and therefore there is not much to show by way of commodities whose market appeal is both factual and documented.

The lack of deliverables lowers the credibility of projects, and more or less does away with the expectancy to get a return on the investments. If there is one top lesson to be retained from the failure of ICOT, and of research projects that IBM had undertaken in the 1980s, this is it.

THE SAD STORY OF EUROPEAN SUPERCOMPUTER INITIATIVES

A fairly common characteristic of government-sponsored projects anywhere in the world is that when the budget on which an organization feeds is assured, the urge to get results diminishes. Eventually, the outcome of an R&D project becomes inversely proportional to the lavishness with which it is funded.

In the case of ICOT, for instance, $50 million per year without a fine grid of milestones has proved a budget rich enough to discourage concrete, well-timed outcomes whose advent should have been a prerequisite to further budgetary allocations. This happens all over the globe and other projects consume much more money than that without a visible output.

Unfortunately with government funding and with R&D financing by large bureaucratic corporations, the lack of results has no penalties associated to it because it is all one big political game. This however does not exclude that there exist background reasons which do not allow deliverables to show through.

One of these reasons is the nature of some government-financed R&D programmes oriented towards pre-competitive research. This limits the implementation domain; the way it has happened, for instance, in the European Union with the Esprit I and Esprit II projects – but not necessarily Esprit III and Esprit IV. The latter presents significant improvements in concepts because it has become business-oriented and market-sensitive.

There is no lack of want to develop a high-performance computing programme in Europe. But there are no results either. A report from a group of leading European scientists admits that although the old continent is far behind in high-performance computing, progress in:

- Understanding, and
- Controlling

complex physical systems is recently on the way of a breakthrough. What seems to be missing is the synergy with technology. The *Report on High-Performance Computing*, requested by the European Commission, links economic and social breakthrough to sophisticated machinery which:

- Will exist by the year 2000, and
- Will revolutionize all fields of science.

Drawn up by high-level industrial and scientific experts, the report of the High-Performance Computing and Networking (HPCN) Advisory Committee concludes that Europe now has a chance of taking the lead in the most advanced computing techniques – but that chance is not necessarily being taken. However what I said about HPCC is just as valid with HPCN. Both are very costly projects which have outlived their purpose.

In February 1994, the European Commission turned to business leaders to co-sponsor a centre for infrastructural research on HPCN. It also asked commercial and industrial organizations to determine:

- Who should invest in the network,
- How it would be run, and
- Who would get access.

In the background of these questions is whether there exists a real consumer and business need for HPCN. The estimated cost of a European-wide information infrastructure is ECU 150 billion ($170 billion), but the Commission will contribute only ECU 5 billion and will additionally fund the Rotterdam-based European Centre for Infrastructure Studies.

In a way paralleling the American HPCC, the Advisory Committee that the European Commission had set up asked for the development of leading-edge information technology, involving the most advanced computing systems interconnected by high-speed networks. Even the name has similarity: HPCN for High-Performance Computing and Networking. (The most ludicrous side of the European proposal is that it requests 10 years funding for a programme subject only to a mid-term review.)

- To assure a competitive position by the start of the 21st century, European scientists urge a very large investment in high-performance computing efforts.
- They call for an annual contribution of ECU 1.1 billion ($1.33 billion) from various community, national and business sources.

Each one of these two bullets has something negative aspect associated to it. In regard to the first bullet, the time of specially designed supercomputer chips is now past. This effort comes 15 years too late. Today's high-performance computers are made of low-cost, off-the-shelf processors.

The other negative is that too much money has been thrown at the different problems and there are no deliverables. The second phase of Esprit, which finished in 1992, represented a European Union investment of more than ECU 3.2 billion ($3.85 billion). But very little came by way of practical results.

Government financing of big science projects consistently proved to be very inefficient. At the core, there is a *political* issue of wrong choices made at the wrong time:

- Much more weight is given to hidden subsidies to industries and university research laboratories, than to innovation.

By and large, government-financed research projects are *good for further funding*, while

- Very little attention is paid to producing market leaders. Yet, this would have been far better both for business and for the university labs.

Spending money alone does not assure success. A clearer vision is necessary in terms of products, markets and software breakthroughs – but, as we have already seen, this is lacking in all government-sponsored high-performance products in America, Europe and Japan.

Acknowledgements

The following organizations and their senior executives participated in the research projects which contributed significantly to the contents of the present book and its documentation. But while valuable input comes from these sources, the evaluation and presentation is the author's own.

UNITED STATES

Bankers Trust

- Dr Carmine VONA, Executive Vice President for Worldwide Technology
- Shalom BRINSY, Senior Vice President Distributed Networks
- Dan W. MUECKE, Vice President, Technology Strategic Planning
- Bob GRAHAM, Vice President, Database Manager

Bankers Trust Plaza, New York, NY 10006

Citibank

- Colin CROOK, Chairman Corporate Technology Committee
- David SCHULTZER, Senior Vice President, Information Technology
- Jim CALDARELLA, Manager, Business Architecture for Global Finance
- Nicholas P. RICHARDS, Database Administrator
- William BRINDLEY, Technology Officer
- Michael R. VEALE, Network Connectivity
- Harriet SCHABES, Corporate Standards
- Leigh REEVE, Technology for Global Finance

399 Park Avenue, New York, NY 10043

Morgan Stanley

- Gary T. GOEHRKE, Managing Director, Information Services
- Guy CHIARELLO, Vice President, Databases
- Robert F. DE YOUNG, Principal, Information Technology

1933 Broadway, New York, NY 10019

- Eileen S. WALLACE, Vice President, Treasury Department
- Jacqueline T. BRODY, Treasury Department

1251 Avenue of the Americas, New York, NY 10020

Goldman Sachs

- Vincent L. AMATULLI, Information Technology, Treasury Department

85 Broad Street, New York, NY 10004

J.J. Kenny Services Inc.

- Thomas E. ZIELINSKI, Chief Information Officer
- Ira KIRSCHNER, Database Administrator, Director of System Programming and of the Data Center

65, Broadway, New York, NY 1006

Merrill Lynch

- Kevin SAWYER, Director of Distributed Computing Services and Executive in Charge of the Mainframe to Client–Server Conversion Process
- Raymond M. DISCO, Treasury/Bank Relations Manager

World Financial Center, South Tower,
New York, NY 10080-6107

Teachers Insurance and Annuity Association/College Retirement Equities Fund (TIAA/CREF)

- Charles S. DVORKIN, Vice President and Chief Technology Officer
- Harry D. PERRIN, Assistant Vice President, Information Technology

730 Third Avenue, New York, NY 10017-3206

Financial Accounting Standards Board

- Halsey G. BULLEN, Project Manager
- Jeannot BLANCHET, Project Manager
- Teri L. LIST, Practice Fellow

401 Merritt 7, Norwalk, CN 06856

Teknekron Software Systems, Inc.

- Vivek RANADIVE, President and CEO
- Robert RECTOR, Senior Vice President, Client Technical Services
- Martin LUTHI, Senior Director, Client Technical Services
- Gerard D. BUGGY, Vice President, Global Financial Sales and Marketing
- Norman CHEUNG, Director, Quantum Leap Group
- Bradley C. RHODE, Vice President, Core Technology Engineering
- Tugrul FIRATLI, Director, Network Consulting Services
- John E. McDOWALL

- Tom JASEK, Director, Market Sheet
- Glenn A. McCOMB, Senior Member of Technical Staff, New Technologies
- Murat K. SONMEZ, Member of Technical Staff
- Murray D. RODE, Member of Technical Staff

530 Lytton Avenue, Suite 301, Palo Alto, CA 94301

Evans and Sutherland

- Les HORWOOD, Director New Business Development
- Mike WALTERMAN, Systems Engineer, Virtual Reality Applications
- Lisa B. HUBER, Software Engineer, 3-Dimensional Programming

600 Komas Drive, PO Box 58700, Salt Lake City, Utah 84158

nCube

- Michael MEIRER, President and Chief Executive Officer
- Craig D. RAMSEY, Senior Vice President, Worldwide Sales
- Ronald J. BUCK, Vice President, Marketing
- Matthew HALL, Director of Software Development

919 East Hillside Blvd, Foster City, CA 94404

Taligent

- Dr Jack GRIMES, Manager, Technology Evaluation
- Catherine JAEGER, Manager, Market Development

10201 N. De Anza Blvd, Cupertino, CA 95014-2233

Visual Numerics

- Don KAINER, Vice President and General Manager
- Joe WEAVER, Vice President OEM/VAR Sales
- Jim PHILLIPS, Director Product Development
- Dr Shawn JAVID, Senior Product Manager
- Dan CLARK, Manager, WAVE Family Products
- Thomas L. WELCH, Marketing Product Manager
- Margaret JOURNEY, Director Administration
- John BEE, Technical Sales Engineer
- Adam ASNES, VDA Sales Executive
- William POTTS, Sales Manager

6230 Lookout Road, Boulder, Colorado 80301

Massachusetts Institute of Technology

- Professor Dr Stuart E. MADNICK, Information Technology and Management Science

- Professor Dr Michael SIEGEL, Information Technology, Sloan School of Management
- Patricia M. McGINNIS, Executive Director, International Financial Services
- Professor Peter J. KEMPTHORNE, Project on Non-Traditional Methods in Financial Analysis
- Dr Alexander M. SAMAROV, Project on Non-Traditional Methods in Financial Analysis
- Robert R. HALPERIN, Executive Director, Center for Coordination Science
- Professor Amar GUPTA, Sloan School of Management
- Professor Jean-Luc VILA, Finance Dept., Sloan School of Management
- Professor Bin ZHOU, Management Science, Sloan School of Management

292 Main Street, Cambridge, MA 02139

- Eric B. SUNDIN, Industrial Liaison Officer
- David L. VERRILL, Senior Liaison Officer, Industrial Liaison Program

Sloan School of Management
50 Memorial Drive, Cambridge, MA 02139

- Henry H. HOUH, Desk Area Network and ViewStation Project, Electrical Engineering and Computer Science
- Dr Henry A. LIEBERMAN, Media Laboratory
- Valerie A. EAMES, Media Laboratory
- Professor Dr Kenneth B. HAASE, Media Arts and Sciences
- Dr David ZELTZER, Virtual Reality Project

Ames St., Cambridge, MA 02139

University of Michigan

- Professor John H. HOLLAND, Electrical Engineering and Computer Science
- Dr Rick L. RIOLO, Systems Researcher, Department of Psychology

Ann Arbor, MI 48109-2103

Santa Fe Institute

- Dr Edward A. KNAPP, President
- Dr L. Mike SIMMONS, Jr., Vice President
- Dr Bruce ABELL, Vice President Finance
- Professor Dr Murray GELL-MANN, Theory of Complexity
- Professor Dr Stuart KAUFFMAN, Models in Biology
- Dr Chris LANGTON, Artificial Life
- Dr John MILLER, Adaptive Computation in Economics
- Dr Blake Le BARON, Non-Traditional Methods in Economics
- Bruce SAWHILL, Virtual Reality

1660 Old Pecos Trail, Santa Fe, NM 87501

School of Engineering and Applied Science, University of California, Los Angeles

- Dean A.R. Frank WAZZAN, School of Engineering and Applied Science
- Professor Richard MUNTZ, Chair, Computer Science Department
- Professor Dr Leonard KLEINROCK, Telecommunications and Networks
- Professor Nicolaos G. ALEXOPOULOS, Electrical Engineering
- Professor Dr Judea PEARL, Cognitive Systems Laboratory
- Professor Dr Walter KARPLUS, Computer Science Department
- Professor Dr Michael G. DYER, Artificial Intelligence Laboratory
- Ms Susan CRUSE, Director of Development and Alumni Affairs
- Joel SHORT, PhD Candidate
- David CHICKERING, PhD Candidate

Westwood Village, Los Angeles, CA 90024

School of Business Administration, University of Southern California

- Dr Bert M. STEECE, Dean of Faculty, School of Business Administration
- Dr Alan ROWE, Professor of Management

Los Angeles, CA 90089-1421

Prediction Company

- Dr J. Doyne FARMER, Director of Development
- Dr Norman H. PACKARD, Director of Research
- Jim McGILL, Managing Director

234 Griffin Street, Santa Fe, NM 87501

Simgraphics Engineering Corp.

- Steve TICE, President
- David J. VERSO, Chief Operating Officer

1137 Huntington Drive, South Passadena,
CA 91030-4563

Nynex Science and Technology, Inc.

- Thomas M. SUPER, Vice President, Research and Development
- Steven CROSS, NYNEX Shuttle Project
- Valerie R. TINGLE, System Analyst
- Melinda CREWS, Public Liaison, NYNEX Labs.

500 Westchester Avenue, White Plains, NY 10604

- John C. FALCO, Sales Manager, NYNEX Systems Marketing
- David J. ANNINO, Account Executive, NYNEX Systems Marketing

100 Church Street, New York, NY 10007

Microsoft

- Mike McGEEHAN, Database Specialist
- Andrew ELLIOTT, Marketing Manager

825, 8th Avenue, New York, NY

Reuters America

- Robert RUSSEL, Senior Vice President
- William A.S. KENNEDY, Vice President
- Buford SMITH, President, Reuters Information Technology
- Richard A. WILLIS, Manager International Systems Design
- M.A. SAYERS, Technical Manager, Central Systems Development
- Alexander FAUST, Manager Financial Products USA (Instantlink and Blend)

40 E. 52nd Street, New York, NY 10022

Oracle Corporation

- Scott MATTHEWS, National Account Manager
- Robert T. FUNK, Senior Systems Specialist
- Joseph M. Di BARTOLOMEO, Systems Specialist
- Dick DAWSON, Systems Specialist

885 Third Avenue, New York, NY 10022

Digital Equipment Corporation

- Mike FISHBEIN, Product Manager, Massively Parallel Systems (MAS-PAR Supercomputer)
- Marco EMRICH, Technology Manager, NAS
- Robert PASSMORE, Technical Manager, Storage Systems
- Mark S. DRESDNER, DEC Marketing Operations

146 Main Street, Maynard, MA 01754 (Meeting held at UBS New York)

Unisys Corporation

- Harvey J. CHIAT, Director Impact Programs
- Manuel LAVIN, Director, Databases
- David A. GOIFFON, Software Engineer

P.O. Box 64942, MS 4463; Saint Paul, MN, 55164–0942 (Meeting held at UBS in New York)

Hewlett-Packard

- Brad WILSON, Product Manager, Commercial Systems
- Vish KRISHNAN, Manager R + D Laboratory

- Samir MATHUR, Open ODB Manager
- Michael GUPTA, Transarc, Tuxedo, Encina Transaction Processing
- Dave WILLIAMS, Industry Account Manager

1911, Pruneridge Avenue, Cupertino, CA 95014

IBM Corporation

- Terry LIFFICK, Software Strategies, Client–Server Architecture
- Paula CAPPELLO, Information Warehouse Framework
- Ed COBBS, Transaction Processing Systems
- Dr Paul WILMS, Connectivity and Interoperability
- Helen ARZU, IBM Santa Teresa Representative
- Dana L. STETSON, Advisory Marketing IBM New York

Santa Teresa Laboratory, 555 Bailey Avenue, San José, CA 95141

UBS Securities

- A. Ramy GOLDSTEIN, Managing Director, Equity Derivative Products

299 Park Avenue, New York, NY 10171–0026

Union Bank of Switzerland

- Dr H. BAUMANN, Director of Logistics, North American Operations
- Dr Ch. GABATHULER, Director, Information Technology
- Hossur SRIKANTAN, Vice President Information Technology Department
- Roy M. DARHIN, Assistant Vice President

299 Part Avenue, New York, NY 10171–0026

UNITED KINGDOM

Barclays Bank

- Peter GOLDEN, Chief Information Officer, Barclays Capitals Markets, Treasury, BZW
- Brandon DAVIES, Director of Financial Engineering
- David J. PARSONS, Director Advanced Technology
- Christine E. IRWIN, Group Information Systems Technology

Murray House, 1 Royal Mint Court, London EC3N 4HH

Bank of England

- Mark LAYCOCK, Banking Supervision Division

Threadneedle Street, London EC2R 8AH

British Telecom

- Dr Alan RUDGE, Deputy Managing Director

BT Centre, 81 Newgate Street, London EC1A 7AJ

Association for Payment Clearing Services (APACS)

- J. Michael WILLIAMSON, Deputy Chief Executive

14 Finsbury Square, London EC2A 1BR

Abbey National Bank

- Mac MILLINGTON, Director of Information Technology

Chalkdell Drive, Shenley Wood, Milton Keynes MK6 6LA

- Anthony W. ELLIOTT, Director of Risk and Credit

Abbey House, Baker Street, London NW1 6XL

Natwest Securities

- Sam B. GIBB, Director of Information Technology,
- Don F. SIMPSON, Director, Global Technology
- Richard E. GIBBS, Director, Equity Derivatives

135 Bishopsgate, London EC2M 3XT

Oracle Corporation

- Mr Geoffrey W. SQUIRE, Executive Vice President, and Chief Executive
- Mr Richard BARKER, Senior Vice President and Director British Research Laboratories
- Mr Giles GODART-BROWN, Senior Support Manager
- Mr Paul A. GOULD, Account Executive

Oracle Park, Bittams Lane, Guildford Rd. Chertsey, Surrey KT16 9RG

Virtual Presence

- Stuart CUPIT, Graphics Engineer

25 Corsham Street, London N1 6DR

Valbecc Object Technology

- Martin FOWLER, Ptech Expert

115 Wilmslow Road, Handforth, Wilmslow, Cheshire SK9 3ER

Credit Suisse Financial Products

- Ross SALINGER, Managing Director

1 Cabot Square, London E14 4QJ

Credit Suisse First Boston

- Geoff J.R. DOUBLEDAY, Executive Director

1 Cabot Square, London E14 4QJ

E.D. & F. Man International

- Brian FUDGE, Funds Division

Sugar Quay, Lower Thames Street, London EC3R 6DU

SCANDINAVIA

Vaerdipapircentralen (VP)

- Mr Jens BACHE, General Manager
- Mrs Aase BLUME, Assistant to the General Manager

61 Helgeshoj Allé, Postbox 20, 2630 Taastrup-Denmark

Swedish Bankers' Association

- Mr Bo GUNNARSSON, Manager, Bank Automation Department
- Mr Gösta FISCHER, Manager, Bank-Owned Financial Companies Department
- Mr Göran AHLBERG, Manager, Credit Market Affairs Department

P.O. Box 7603, 10394 Stockholm-Sweden

Skandinaviska Enskilda Banken

- Mr Lars ISACSSON, Treasurer
- Mr Urban JANELD, Executive Vice President Finance and IT
- Mr Mats ANDERSSON, Director of Computers and Communications
- Mr Gösta OLAVI, Manager SEB Data/Koncern Data

2 Sergels Torg, 10640 Stockholm-Sweden

Securum AB

- Mr Anders NYREN, Director of Finance and Accounting
- Mr John LUNDGREN, Manager of IT

38 Regeringsg, 5 tr., 10398 Stockholm-Sweden

Sveatornet AB of the Swedish Savings Banks

- Mr Gunar M. CARLSSON, General Manager

(Meeting at Swedish Bankers' Association)

Mandamus AB of the Swedish Agricultural Banks

- Mrs Marie MARTINSSON, Credit Department

(Meeting at Swedish Bankers' Association)

Handelsbanken

- Mr Janeric SUNDIN, Manager, Securities Department
- Mr Jan ARONSON, Assistant Manager, Securities Department

(Meeting at Swedish Bankers' Association)

Gota Banken

- Mr JOHANNSSON, Credit Department

(Meeting at Swedish Bankers' Association)

Irdem AB

- Mr Gian MEDRI, Former Director of Research at Nordbanken

19 Flintlasvagen, 19154 Sollentuna-Sweden

AUSTRIA

Creditanstalt Bankverein

- Dr Wolfgang G. LICHTL, Director of Foreign Exchange and Money Markets
- Dr Johann STROBL, Manager, Financial Analysis for Treasury Operations

3. Julius Tandler-Platz, 1090 Vienna

Bank Austria

- Dr Peter FISCHER, Director of Treasury
- Peter GABRIEL, Deputy General Manager, Trading
- Konrad SCHCATE, Manager, Financial Engineering

2, Am Hof, 1010 Vienna

Association of Austrian Banks and Bankers

- Dr Fritz DIWOK, Secretary General

11, Boersengasse, 1013 Vienna

Aktiengesellschaft Fuer Bauwesen

- Dr Josef Fritz, General Manager

2, Lothringenstrasse, 1041 Vienna

Management Data of Creditanstalt

- Ing. Guenther REINDL, Vice President, International Banking Software
- Ing. Franz NECAS, Project Manager, RICOS
- Mag. Nikolas GOETZ, Product Manager, RICOS

21–25 Althanstrasse, 1090 Vienna

GERMANY

Deutsche Bundesbank

- Eckhard OECHLER, Director of Bank Supervision and Legal Matters

14, Wilhelm Epstein Strasse, D-6000 Frankfurt 50

Deutsche Bank

- Peter GERARD, Executive Vice President, Organization and Information Technology
- Hermann SEILER, Senior Vice President, Investment Banking and Foreign Exchange Systems
- Dr KUHN, Investment Banking and Foreign Exchange Systems
- Dr Stefan KOLB, Organization and Technological Development

12, Koelner Strasse, D-6236 Eschborn

Dresdner Bank

- Dr Karsten WOHLENBERG, Project Leader Risk Management, Simulation and Analytics Task Force Financial Division
- Hans-Peter LEISTEN, Mathematician
- Susanne LOESKEN, Organization and IT Department

43, Mainzer Landstrasse, D-6000 Frankfurt

Commerzbank

- Helmut HOPPE, Director Organization and Information Technology
- Hermann LENZ, Director Controllership, Internal Accounting and Management Accounting
- Harald LUX, Manager Organization and Information Technology
- Waldemar NICKEL, Manager Systems Planning

155, Mainzer Landstrasse, D-60261 Frankfurt

Deutscher Sparkassen und Giroverband

- Manfred KRUEGER, Division Manager, Card Strategy

4 Simrockstrasse, D-5300 Bonn 1 (Telephone interview from Frankfurt)

ABN-AMRO (Holland)

- Mr SCHILDER, Organization and Information Technology

(Telephone interview from Frankfurt)

Media Systems

- Bertram ANDERER, Director

6, Goethestrasse, D-7500 Karlsruhe

Fraunhofer Institute for Computer Graphics

- Dr. Ing. Martin GOEBEL
- Wolfgang FELBER

7, Wilhelminerstrasse, D-6100 Darmstadt

GMD First – Research Institute for Computer Architecture, Software Technology and Graphics

- Professor Dr Ing. Wolfgang K. GILOI General Manager
- Dr BEHR, Administrative Director
- Dr Ulrich BRUENING, Chief Designer
- Dr Joerg NOLTE, Designer of Parallel Operating Systems Software
- Dr Matthias KESSLER, Parallel Languages and Parallel Compilers
- Dr Friedrich W. SCHROER, New Programming Paradigms
- Dr Thomas LUX, Fluid Dynamics, Weather Prediction and Pollution Control Project

5, Rudower Chaussee, D-1199 Berlin

Siemens Nixdorf

- Wolfgang WEISS, Director of Banking Industry Office
- Bert KIRSCHBAUM, Manager, Dresdner Bank Project
- Mark MILLER, Manager Neural Networks Project for UBS and German banks
- Andrea VONERDEN, Business Management Department

27, Lyoner Strasse, D-6000 Frankfurt 71

UBS Germany

- H.-H. v. SCHELIHA, Director Organization and Information Technology
- Georg SUDHAUS, Manager IT for Trading Systems
- Marco BRACCO, Trader
- Jaap VAN HARTEN, Trader

52, Bleichstrasse, D-6000 Frankfurt 1

SWITZERLAND

Bank for International Settlements

- Claude SIVY, Director, Controllership and Operational Security
- Frederik C. MUSCH, Secretary General, Basel Committee on Banking Supervision

2 Centralbankplatz, Basel

Ciba-Geigy AG

- Stefan JANOVJAK, Divisional Information Manager
- Natalie PAPEZIK, Information Architect

Ciba-Geigy, R-1045, 5.19, 4002 Basle

Eurodis

- Albert MUELLER, Director
- Beat ERZER, Marketing Manager
- B. PEDRAZZINI, Systems Engineer
- Reto ALBERTINI, Sales Engineer

Bahnhofstrasse 58/60, CH-8105 Regensdorf

Olsen and Associates

- Dr Richard OLSEN, President

233, Seefeldstrasse, Zurich

JAPAN

Bank of Japan

- Harry TOYAMA, Councel and Chief Manager, Credit and Market Management Department
- Akira IEDA, Credit and Market Management Department

2–1–1, Kongoku-Cho, Nihonbashi, Chuo-ku, Tokyo 103

Dai-Ichi Kangyo Bank

- Shunsuke NAKASUJI, General Manager and Director, Information Technology Division
- Seiichi HASEGAWA, Manager International Systems Group
- Takahiro SEKIZAWA, International Systems Group
- Yukio HISATOMI, Manager Systems Planning Group
- Shigeaki TOGAWA, Systems Planning Group

13–3, Shibuya, 2-Chome, Shibuya-ku, Tokyo 150

Fuji Bank

- Hideo TANAKA, General Manager Systems Planning Division
- Toshihiko UZAKI, Manager Systems Planning Division
- Takakazu IMAI, Systems Planning Division

Otemachi Financial Center, 1–5–4 Otemachi, Chiyoda-ku, Tokyo

Mitsubishi Bank

- Akira WATANABE, General Manager, Derivative Products
- Akira TOWATARI, Manager, Strategic Planning and Administration, Derivative Products
- Takehito NEMOTO, Chief Manager, Systems Development Division
- Nobuyuki YAMADA, Systems Development Division
- Haruhiko SUZUKI, Systems Development Division

7–1, Marunouchi, 2-chome, Chiyoda-ku, Tokyo 100

Nomura Research Institute

- Tomio ARAI, Director, Systems Science Department
- Tomoyuki OHTA, Director, Financial Engineering Group
- Tomohiko HIRUTA, Manager, I-STAR Systems Services

9–1, Nihonbashi, 1-Chome, Chuo-ku, Tokyo 103

Mitsubishi Trust and Banking

- Nobuyuki TANAKA, General Manager, Systems Planning Division
- Terufumi KAGE, Consultant Systems Planning Division

9–8 Kohnan, 2-Chome, Minato-ku, Tokyo 108

Sakura Bank

- Nobuo IHARA, Senior Vice President and General Manager, Systems Development Office VIII
- Hisao KATAYAMA, Senior Vice President and General Manager, System Development Office VII
- Toshihiko EDA, Senior Systems Engineer, Systems Development Division

4–2, Kami-Osahi, 4-Chome, Shinagawa-ku, Tokyo 141

Sanyo Securities

- Yuji OZAWA, Director, Systems Planning Department
- K. TOYAMA, Systems Planning Department

1–8–1, Nihonbashi, Kayabacho, Chuo-ku, Tokyo 103

Center for Financial Industry Information Systems (FISC)

- Shighehisa HATTORI, Executive Director
- Kiyoshi KUMATA, Manager, Research Division II

16th Floor, Ark Mori Building, 12–32, 1-Chome Akasaka, Minato-ku, Tokyo 107

Laboratory for International Fuzzy Engineering Research (LIFE)

- Professor Dr Toshiro TERANO, Executive Director
- Dr Anca L. RALESCU, Assistant Director
- Shunichi TANI, Fuzzy Control Project Leader Siber Hegner Building, 89–1 Yamashita-Cho, Naka-ku, Yokohama-shi 231

Real World Computing Partnership (RWC)

- Dr Junichi SHUMADA, General Manager of RWC
- Hajime IRISAWA, Executive Director

Tsukuba Mitsui Building, 1–6–1 Takezono,
Tsukuba-shi, Ibarahi 305

Tokyo University

- Professor Dr Michitaka HIROSE, Dept. of Mechano-Informatics, Faculty of Engineering
- Dr Kensuke YOKOYAMA, Virtual Reality Project

3–1, 7-Chome, Hongo Bunkyo-ku, Tokyo 113

Tokyo International University

- Professor Dr Yoshiro KURATANI

9–1–7–528, Akasaka, Minato-ku, Tokyo 107

Japan Electronic Directory Research Institute

- Dr Toshio YOKOI, General Manager

Mita-Kokusai Building – Annex, 4–28 Mita, 1-Chome, Minato-ku, Tokyo 108

Mitsubishi Research Institute (MRI)

- Masayuki FUJITA, Manager, Strategic Information Systems Dept.
- Hideyuki MORITA, Senior Research Associate, Information Science Dept.
- Akio SATO, Research Associate, Information Science Dept.

ARCO Tower, 8–1 Shimomeguro, 1-Chome, Meguro-ku, Tokyo 153

NTT Software

- Dr Fukuya ISHINO, Senior Vice President

223–1 Yamashita-Cho, Naka-ku, Yokohama 231

Ryoshin Systems (Systems Developer Fully Owned by Mitsubishi Trust)

- Takewo YUWI, Vice President, Technical Research and Development

9–8 Kohman, 2-Chome, Minato-ku, Tokyo 108

Sanyo Software Services

- Fumio SATO, General Manager, Sales Department 2

Kanayama Building, 1–2–12 Shinkawa, Chuo-ku, Tokyo 104

Fujitsu Research Institute

- Dr Masuteru SEKIGUCHI, Member of the Board and Director of R+D
- Takao SAITO, Director of the Parallel Computing Research Center
- Dr Hiroyasu ITOH, R+D Department
- Katsuto KONDO, R+D Department
- Satoshi HAMAYA, Information Systems and Economics

9–3 Nakase, 1-Chome, Mihama-ku, Chiba-City 261

NEC

- Kotaro NAMBA, Senior Researcher, NEC Planning Research
- Dr Toshiyuki NAKATA, Manager, Computer System Research Laboratory
- Asao KANEKO, Computer System Research Laboratory

3–13–12 Mita, Minato-ku, Tokyo 108

Toshiba

- Dr Makoto IHARA, Manager Workstation Product Planning and Technical Support Dept.
- Emi NAKAMURA, Analyst Financial Applications Dept.
- Joshikiyo NAKAMURA, Financial Sales Manager
- Minami ARAI, Deputy Manager, Workstation Systems Division

1–1, Shibaura, 1-Chome, Minato-ku, Tokyo 105

Microsoft

- James LALONDE, Multinational Account Manager, Large Accounts Sales Dept.

Sasazuka NA Bldg, 50–1 Sasazuka, 1-Chome, Shibuya-ku, Tokyo 151

Apple Technology

- Dr Tsutomu KOBAYASHI, President

25 Mori Bldg, 1–4–30 Roppongi, Minato-ku, Tokyo 106

Digital Equipment Japan

- Roshio ISHII, Account Manager, Financial Sales Unit 1

2–1 Kamiogi, 1-Chome, Suginamiku, Tokyo 167

UBS Japan

- Dr Peter BRUTSCHE, Executive Vice President and Chief Manager
- Gary P. EIDAM, First Vice President, Regional Head of Technology
- Charles UNDERWOOD, Vice President, Head of Technical Architecture and Strategy
- Masaki UTSUNOMIYA, MANAGER, IT Production Facilities

Yurakucho Building 2F, 1–10–1 Yurakucho, Chiyoda-ku, Tokyo 100

Bibliography

Business Week (1993) 'American Investors are Reinventing Themselves', 18 January, p. 48.

Business Week (1994) 11 April.

Business Week (1994) 21 November.

Business Week (1995) 30 January.

Carroll, Paul (1993) *Big Blues: The Unmasking of IBM* (New York: Crown).

Chorafas, Dimitris N. (1960) *Statistical Processes and Reliability Engineering* (Princeton, NJ: Van Nostrand).

— (1989) *System Architecture and System Design* (New York: McGraw-Hill).

— (1991) *The Handbook of Data Communications and Computer Networks*, 2nd edn (New York: McGraw-Hill/Tab Books).

— (1992) *The New Information Technologies – A Practitioner's Guide* (New York: Van Nostrand Reinhold).

— (1994) *Chaos Theory in the Financial Markets* (Chicago: Probus-Irwin).

— (1994a) *Intelligent Multimedia Databases* (Englewood Cliffs, NJ: Prentice Hall).

— (1994b) *Beyond LANs, Client–Server Computing* (New York: McGraw-Hill).

— (1995) *Derivative Financial Instruments* (London & Dublin: Lafferty Publications).

— (1995a) *Managing Derivatives Risk* (Chicago: Probus-Irwin).

— (1996) 'How to Understand and Use Mathematics for Derivatives' Vol. 2 (London: Euromoney)

— (1996a) *Understanding Agent Technology* (London & New York: International Thompson Publishing).

Chorafas, Dimitris M. and Steinmann, Heinrich (1989) *Supercomputers* (New York: McGraw-Hill).

— (1990) *Intelligent Networks, Telecommunications Solutions for the 1990s* (Boca Raton, FL: CRC Press).

— (1993) *Solutions for Networked Databases* (San Diego, CA: Academic Press).

— (1993a) *Object-Oriented Databases* (Englewood Cliffs, NJ: Prentice Hall).

— (1994) *Off-Balance Sheet Financial Instruments* (Chicago: Probus-Irwin).

— (1994a) *Database Mining* (London & Dublin: Lafferty).

— (1995) *Virtual Reality – Practical Applications in Business and Industry* (Englewood Cliffs, NJ: Prentice Hall).

Communications of the ACM (1991) vol. 34, November.

Communications Week International (1995) 7 August.

Daily Pilot (1982) 20 June.

International Herald Tribune (1994) 9 May.

Index